APARTMENT DESIGN

APARTMENT DESIGN

CHRIS VAN UFFELEN

BRAUN

CONTENTS

Der Karl-Marx-Hof in Wien.

PREFACE

Apartment buildings are by far the most important urban building block. Although museums, train stations, and libraries are what we think of when we think of individual cities, these solitary buildings only ever characterize a square or a street. The majority of places and blocks consist of residential buildings, followed by office buildings. While multi-story residential housing is found in smaller cities, the number and height of these buildings increases with population density and urbanization.

Historically, it was also the housing of the plebs urbana that characterized the city; even if townhouses or palaces dominate our idea of urban living, it was the general population that had a greater share in the social fabric and therefore required more living space overall. In Ostia Antica, the harbor of ancient Rome, for example, the so-called insulae with five or six floors have been preserved. The simple homes of the plebs urbana, the cenacula, usually had one or two rooms and an area of about 20 square meters. Commercial units were located on the ground floor, and the more stairs you had to climb to get to your apartment, the cheaper it was. This remained the case until the fall-proof elevator became a standard feature of multi-story residential buildings in the second half of the 19th century.

The Industrial Revolution led to a large increase in urban population as large numbers of workers moved to the cities, which were now increasingly free of fortifications and provided job opportunities in industry. At the same time, part of the historical urban population - the aristocracy and the wealthy bourgeoisie - moved to the countryside, away from the factories. Their vacant flats were subdivided into Maisson Mixte, where different classes all lived under one roof. Again, the most expensive apartments were at the bottom, on the Bel Etage, and social status declined as one moved up. This social segmentation was then deliberately built - think of Georges-Eugène Baron Haussmann's boulevards in Paris. The housing situation of the poorer population remained challenging for a long time, and numerous neighborhoods for the less privileged were built in the late 19th century.

The risk of disease transmission in these neighborhoods prompted philanthropic members of the middle class to seek improvement of the housing situation. However, it was the cities of Vienna and Amsterdam, both of which were

Rue de Rivoli in Paris.

governed early on by social democrats, that made mass housing possible in the early 20th century. Both set high aesthetic standards, as they were supposed to be palaces for the workers. It was not until the New Architecture movement of the 1920s that perimeter block development was abandoned in favor of healthier terraced housing in many places: Light and air should prevent tuberculosis from spreading easily.

The fascination with skyscrapers came from America at the beginning of the 20th century, although over there skyscrapers were only used for offices. In the meantime, the residential skyscraper has returned to America as a particularly prestigious form of housing.

The period after the Second World War saw further development of these types. Le Corbusier added interior pedestrian zones to the high terraced buildings and opened up the roof as a lounge area (Unité d'Habitation, 1952), while satellite towns (Bijlmer in Amsterdam by Siegfried Nassuth, 1966, or in the greater Paris area) emerged in many places, following the idea of Ebenezer Howard's Garden City (1902). New floor plans emerged, especially the open plan from kitchen to dining area to living room - often without the corridors that had previously given a clear structure to housing. Maisonettes on pergolas – although a development of the years between the two world wars – became popular. Panel construction – no longer exclusive to communist countries – industrialized the building process and introduced the desire for a new postmodern comfort. Meanwhile, the wealthier classes have long since returned to the city, and even if the townhouse plays an important role for them, it is still the building types that existed at that time that characterize housing today: The perimeter block in vertical orientation as a result of the development of the historic city, in horizontal alignment as a result of the early social reformers, the often high terraced buildings of classical modernism, and the high-rise residential building as a solitaire or in groups.

Postmodernism was once again critical of the floor plan without corridors, of row houses, and of city life itself. But the urban exodus of the late 20th century has now come to an end. In 1950, barely one-third of the world's population lived in cities; by 2023, it will be 57 percent, and by 2050, it will probably be 68 percent. Today, more and more people are enjoying the diverse opportunities offered by urban areas – including their growing green spaces – and the freedom of living in the city. It was said in the Middle Ages that „city air makes you free,“ and it was meant literally: in the city, people were freed from serfdom and economic dependence. Today it is true in a figurative sense. But cities still have problems: climatic (e. g. high temperatures), economic (rents), logistical (supply) and social (the division between rich and poor). Architects and developers in this volume also have to deal with some of these issues. The answers are manifold, and Apartment Building Architecture shows housing for rich and poor, strategies against climate change, and repeatedly asks how the individual urban building block can be integrated into the fabric of the city.

VALLEY

AMSTERDAM, THE NETHERLANDS

The dramatic, geologically inspired, plant-covered Valley stands out in Amsterdam's Zuidas with its three towers of 67, 81, and 100 meters and canti-levered apartments. The building has multiple faces. Its outer shell of smooth mirrored glass blends with the business district, while inside, it appears as though the glass has crumbled away, revealing craggy rock faces of natural stone and greenery. Valley distinguishes itself from other buildings by combining offices, shops, catering, cultural facilities, and apartments in one structure, offering the occupants many conveniences in the vicinity. Unlike the closed-off buildings in Zuidas, its green valley, winding between the towers on the fourth and fifth floors, is accessible via two external stone staircases. More than 220 species of plants are housed in planters throughout, maintained by an irrigation system and façade gardeners. Valley also exceeds sustainability standards, with a BREEAM-NL Excellent certification and high energy performance.

Valley, standing out of the surrounding area.

ARCHITECTS: MVRDV /// **COMPLETION:** 2022 /// **CLIENT:** EDGE /// **INVESTOR:** PRIVATE /// **LANDSCAPE DESIGN:** PIET OUDOULF, DELTAVORM GROEP /// **INTERIOR DESIGN:** HEYLIGERS ARCHITECTS /// **CONTEXT:** URBAN /// **GROSS FLOOR AREA:** 75,000 M² /// **NUMBER OF APARTMENTS:** 200 /// **MAIN SUSTAINABLE FEATURES:** BREAM-NL EXCELLENT, GPR SCORE 8 /// **PHOTOS:** DARIA SCAGLIOLA / WWW.SCAGLIOLABRAKKEE.NL (10, L.), OSSIP VAN DUIVENBODE / WWW.OSSIP.NL

Section.

The cantilevering apartment and terrace arrangement and the planters within the façade.

Views from within the valley, an apartment, and the first floor business district.

The building at dusk from different perspectives.

First floor plan.

A street level view of the spectacular rising volume of the three towers and their cantilevering elements.

Front view with both towers and a ground level view of the tower.

MONTE-LENA BY TARRAF

SÃO JOSÉ DO RIO PRETO, BRAZIL

Montelena by Tarraf, designed by Königsberger Vannucchi, is a high-end development in São José do Rio Preto. It features two towers on a 3,150-square-meter plot. Áurea has one unit per floor, while Rubra has two. The 33 units, ranging from 256 to 343 square meters, each include four suites and are designed to maximize natural light and ventilation. This approach saves energy for lighting and temperature control, while creating a more comfortable environment for residents. The towers' façades feature an original, fluid, and geometric aesthetic. This gives the buildings a unique appearance in the cityscape. Strategically positioned, the buildings create two distinct spaces: an inviting front boulevard open to the city and a secluded leisure area at the rear of the site. This offers residents the opportunity to connect with others or enjoy a more private outdoor space.

Oblique perspective of the buildings dynamic façade under the sunset light.

ARCHITECTS: KÖNIGSBERGER VANNUCHI /// **COMPLETION:** 2022 /// **CLIENT:** TARRAF /// **LANDSCAPE DESIGN:** LUCIANO FIASCHI ARQUITETURA PAISAGÍSTICA /// **INVESTOR:** PRIVATE /// **INTERIOR DESIGN:** S. C. ARQUITETURA E DESIGN /// **CONTEXT:** URBAN /// **GROSS FLOOR AREA:** 16,997 M² /// **NUMBER OF APARTMENTS:** 33 /// **MAIN SUSTAINABLE FEATURES:** INTELLIGENT ENERGY MANAGEMENT SYSTEMS, LED LIGHTING SOLUTIONS, SMART ELEVATORS AND AIR CONDITIONING SYSTEMS, ELECTRIC VEHICLE CHARGING STATIONS /// **PHOTOS:** PEDRO VANNUCHI / WWW.PEDROVANNUCCHI.COM

Detailed close up of geometric terraces and a drone view from above capturing the alternation of terraces and the entrance to the building.

Typical floor plan.

Details of the geometric terraces, which emphasize the iconic façade of the towers from different points of view.

Ground floor plan.

KANNAI BLADE RESIDENCE

YOKOHAMA, JAPAN

Kannai Blade Residence is an apartment building covered with thin concrete panels that create light and shadow across the façades. Special attention was paid to making the building's façade as unique and characterful as the surrounding buildings. The accentuated concrete floor slabs reflect the horizontal linearity of the area, while the sword-like thin concrete panels break the monotony and introduce playfulness. These panels are made of hybrid prestressed concrete reinforced with carbon cables. These thin panels maintain privacy while giving the building an open, airy feel, connecting the occupants to the outside world. The entrance, unusually large for a one-room apartment block, is illuminated by HPC panels with embedded circular acrylic shapes, resembling a starry sky. The design breaks away from traditional façade norms, drawing inspiration from post-war architecture to create a modern building that blends in with its urban surroundings.

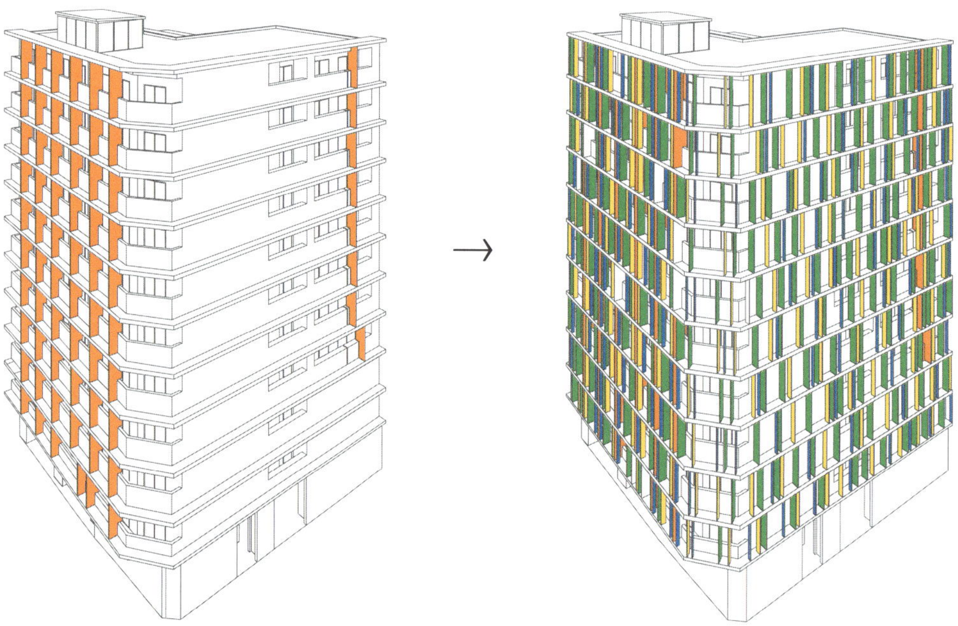

The buildings unique façade and exterior from different perspectives.

Diagram of the façade.

ARCHITECTS: KEY OPERATION INC./ARCHITECTS /// **COMPLETION:** 2021 /// **CLIENT:** TOHSHIN PARTNERS /// **INVESTOR:** PRIVATE /// **CONTEXT:** CITY CENTER /// **GROSS FLOOR AREA:** 3,251 M² /// **NUMBER OF APARTMENTS:** 94 /// **PHOTOS:** TOSHIYUKI YANO

NITTSU

シニアホテル横浜
消火栓
FireBu

Views from a balcony and of the interior of an apartment as well as of the entrance area.

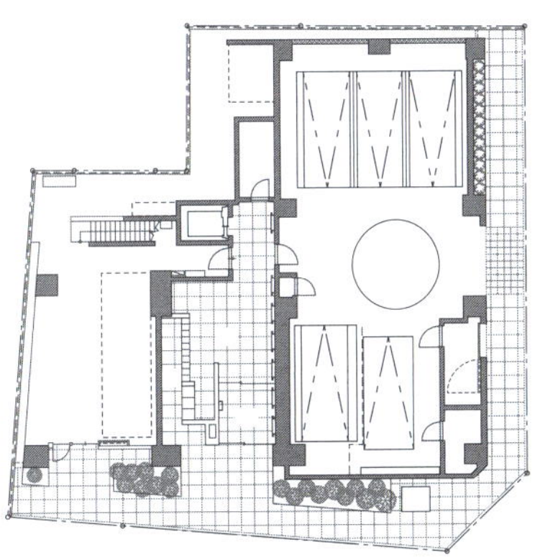

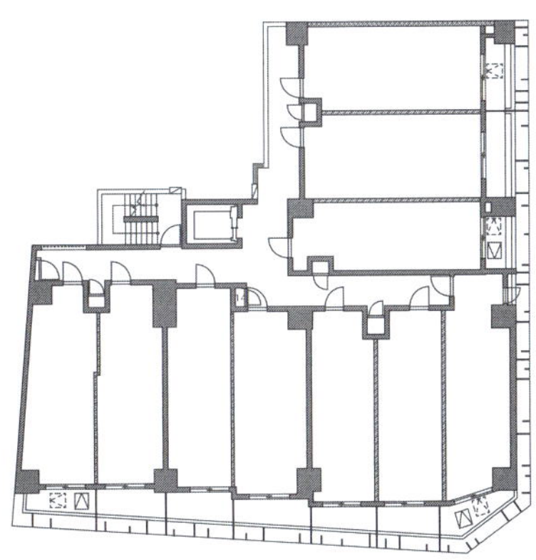

Street level views of the building, showing its façade and contextualizing the surroundings.

Typical floor plan.

MEILLER GÄRTEN

MUNICH, GERMANY

Based on the master plan by rpmp Architekten+Stadtplaner and Reiner Schmidt Landschaftsarchitekten, a residential quarter with around 650 residential units has been built on the northern part of the former Meiller site in Munich-Moosach. The construction site by steidle architekten bordered the existing Meiller factory buildings to the south. The large-scale attitude of historic industrial buildings was deliberately sought at this location. Clinker brick façades on the streets refer to the façades of the former and current production halls. All façades are given a superordinate grid, which basically reflects the respective residential structure and strengthens the building as a large form. The special internal structure, in which separately accessed "house within a house" types alternate with one-to-two-room apartments and maisonette apartments with open-plan living and dining areas, also characterizes the external appearance. This is also pursued in a more differentiated way in the point buildings.

A view of Meiller Gärten's clinker façade, referencing typical factory designs.

The balconies, loggias, and private front yards on the ground floor.

ARCHITECTS: STEIDLE ARCHITEKTEN /// **COMPLETION:** 2021 /// **CLIENT:** RATHGEBER AG MÜNCHEN /// **INVESTOR:** PRIVATE /// **OPEN SPACE DESIGN:** REINER SCHMIDT LANDSCHAFTSARCHITEKTEN, MÜNCHEN /// **CONTEXT:** URBAN /// **GROSS FLOOR AREA:** 15,000 M^2 /// **NUMBER OF APARTMENTS:** 149 /// **MAIN SUSTAINABLE FEATURES:** PV SYSTEM FOR THE HEAT PUMPS, GREEN ROOFS INCLUDING TREES, MOBILITY SERVICES, LOW DENSITY OF DEVELOPMENT TO AVOID THE HEAT ISLAND EFFECT /// **PHOTOS:** STEFAN MÜLLER-NAUMANN / WWW.ARCHITEKTUR-FOTOGRAF-MUENCHEN.DE

Cross section.

View of the point buildings highlighting its openness to natural light and ventilation.

The various entrance areas from the street and a typical apartment view with a loggia.

Close up of Walan's façade, with colored screens protecting from sun and heat gain.

ARCHITECTS: BUREAU^PROBERTS /// **COMPLETION:** 2018 /// **CLIENT:** GBW INVESTMENTS /// **INVESTOR:** PRIVATE /// **LANDSCAPE DESIGN:** RPS /// **CONTEXT:** CITY CENTER /// **GROSS FLOOR AREA:** 8,000 M² /// **NUMBER OF APARTMENTS:** 14 /// **PHOTOS:** CHRISTOPHER FREDERICK JONES / WWW.CFJPHOTO.COM.AU

WALAN

BRISBANE, AUSTRALIA

Walan is designed for subtropical living in Brisbane with a series of whole floor apartment homes stacked to form a high-rise structure that takes a prominent place in the city and riverscape of Brisbane. The development is embedded in the Kangaroo Point peninsula with the fabric and design of the building intrinsically connected to its natural and built environment. The homes on each floor are open to their aspect and protected by screens that provide privacy and shade. The screens are articulated and colored to form an organic fabric that reflects the adjacent rock faces of Kangaroo Point Cliffs. The resulting form creates an ambiguous sculptural presence that reimagines the subtropical, whole-floor apartment tower while celebrating Brisbane's vernacular architecture traditions. These relationships to climate and context ultimately provide a high level of amenity that rivals those found in a detached dwelling.

Street level and river views of Walan's articulating form.

Walan's foyer with glass and cantilevering volume above as well as a view from the east-facing media room of a typical Walan apartment.

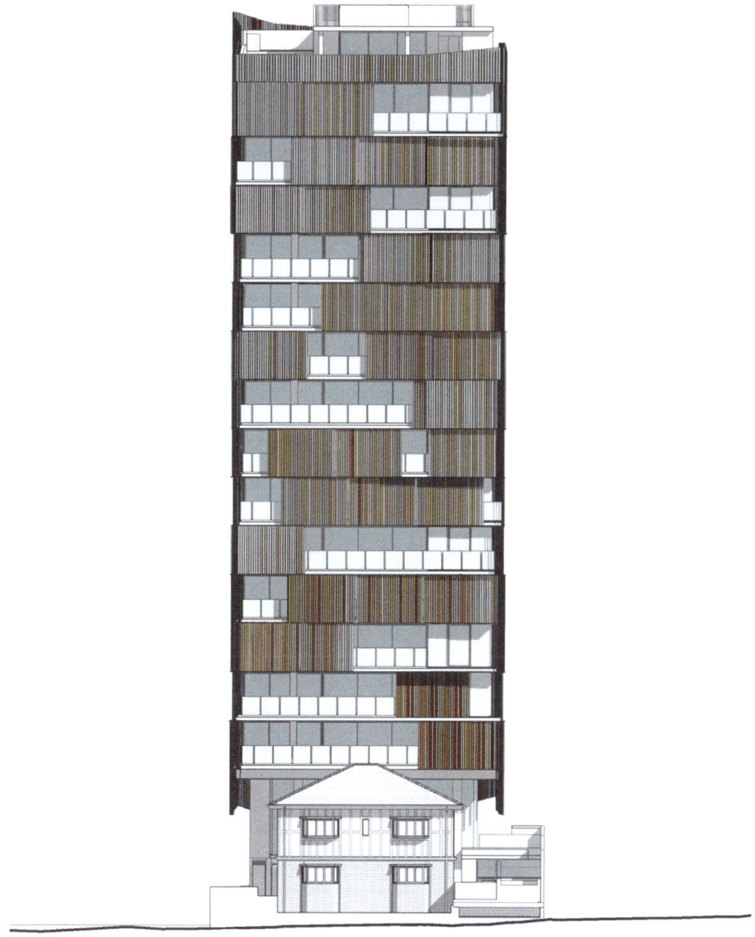

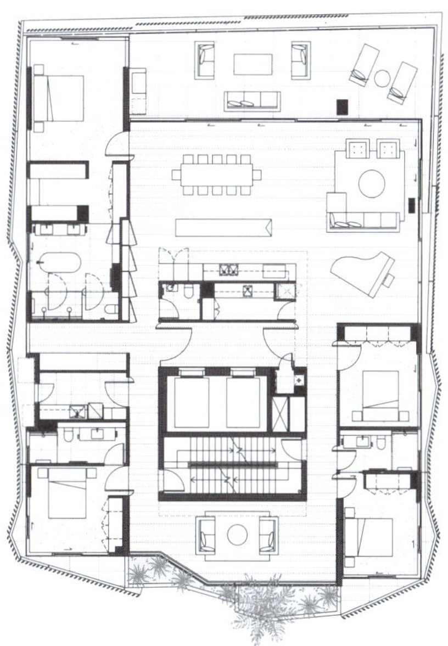

Walan's rising above the river.

West elevation.
Typical floor plan.

LONDON GASHOLDERS

LONDON, UNITED KINGDOM

King's Cross is the largest urban development in Europe. One of the most distinctive features to be retained is the trio of cast iron gasholder guide frames. WilkinsonEyre's design includes three drums of varying heights, reminiscent of the original gasholders. A fourth drum shape forms an open courtyard at the intersection of the structures. Old and new are brought together in the design to create a dynamic counterpoint. The guide frames have a heavy industrial look while the interiors are light and intricate. The circular nature of the buildings provides ample natural daylight. The development comprises 145 apartments, a private gym, spa, business lounge and entertainment suite. All apartments have access from a central courtyard, with each having an atrium and core, linked by walkways with a reflecting water feature. The cladding consists of vertical modular steel and glass panels, textured with an adjustable veil of shutters. The Gasholders project combines modern living with innovative technology, while celebrating the unique character of the Triplets.

The gasholders with different heights.

ARCHITECTS: WILKINSONEYRE ARCHITECTS /// **COMPLETION:** 2018 /// **CLIENT:** ARGENT LLP /// **INVESTOR:** PRIVATE /// **LANDSCAPE DESIGN:** DAN PEARSON STUDIO /// **INTERIOR DESIGN:** JONATHAN TUCKEY DESIGN /// **ORIGINAL BUILDING:** GASWORKS, 1867 /// **CONTEXT:** URBAN /// **GROSS FLOOR AREA:** 23,950 M² /// **NUMBER OF APARTMENTS:** 145 /// **MAIN SUSTAINABLE FEATURES:** COMPREHENSIVE CARBON REDUCTION STRATEGY LINKED TO CCHP SCHEME /// **PHOTOS:** PETER LANDERS / WWW.PETERLANDERS.NET

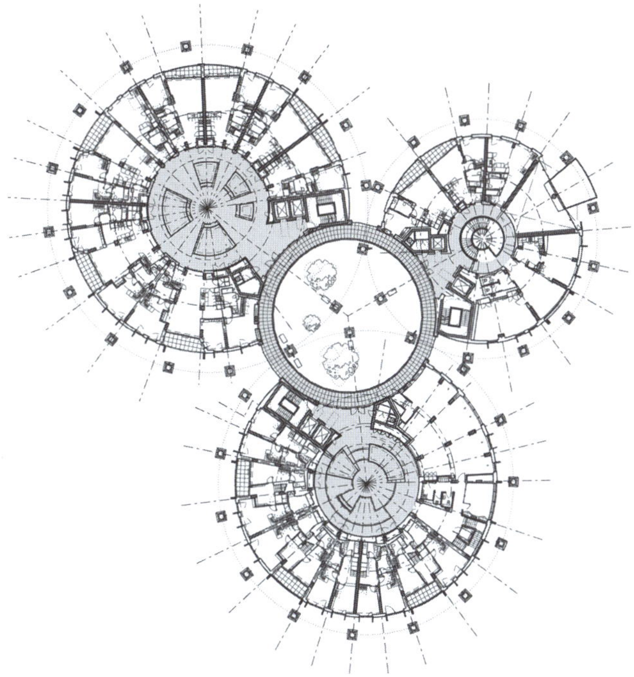

First floor plan.

The exterior façade and public area.

Views of the interiors atrium and an apartment balcony.

Detailed views of the exterior and interior courtyard.

A rear view of the original building that shows the imposing design of the new volume.

ARCHITECTS: AI STUDIO /// **COMPLETION:** 2021 /// **CLIENT:** VESTA DEVELOPMENT /// **ORIGINAL BUILDING:** TELEPHONE EXCHANGE BUILDING, V. PATEK, 1927 /// **CONTEXT:** URBAN /// **GROSS FLOOR AREA:** 18,600 M² /// **NUMBER OF APARTMENTS:** 127 /// **PHOTOS:** DMITRIY CHEBANENKO / WWW.CHEBANENKO.COM

TATLIN APARTMENTS

MOSCOW, RUSSIA

Tatlin Apartments is a unique project in the Moscow cityscape. It combines a classic example of Russian industrial constructivism with a modern apartment complex. As the original building fell into a state of decay, the architects wanted to preserve as much of it as possible, even though it had no formal protected status. Restoration of the historic façade returned the building to its original aesthetic with its distinctive gray and red color. The rear of the building was designed to accommodate two lobbies and elevators that provide access to the apartments above. These apartments, as well as a public sky garden, are located in a strikingly modern building with a triangular cantilevered façade over the historic building. The façade of the cantilevered volumes is an iconic contemporary design, a grid of large metal components and stained glass that intentionally contrasts with the original building while also complementing it.

Front view of the original building and side view of the new volume placed on the old building.

A street level view of Tatlin Apartments to provide context of its surroundings.

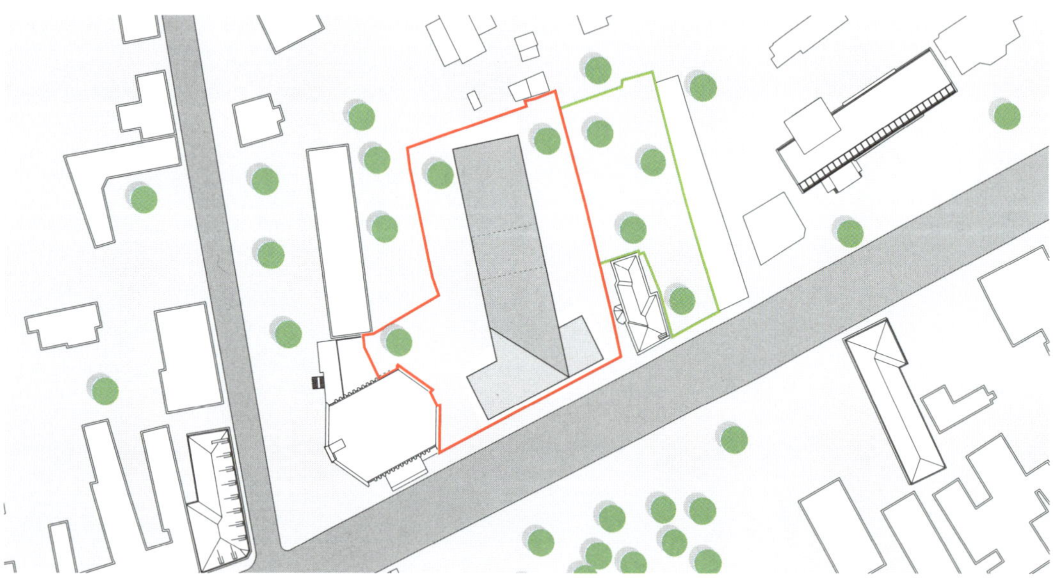

Site plan.

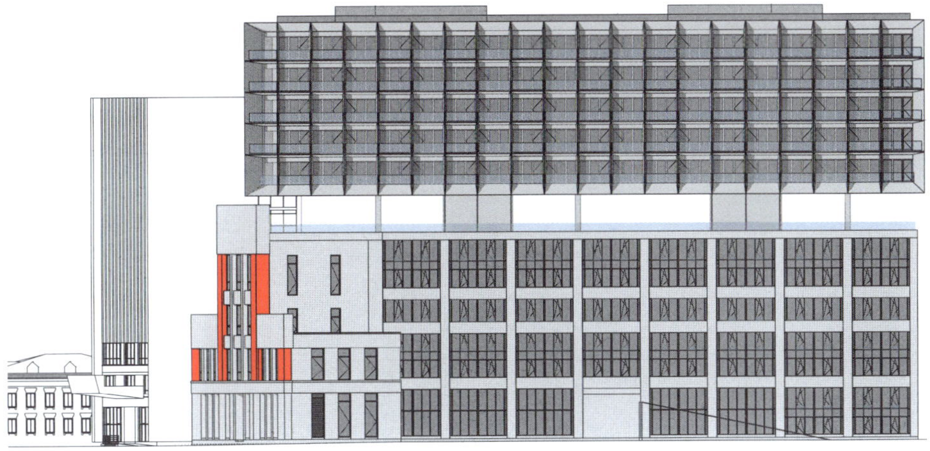

A front view of the triangularly tapered façade and a close up of the modern grid façade design.

Elevation.

БАКУНИНСКАЯ
УЛИЦА
5

РЕСТОРАН
3-е КОЛЬЦО
А103 РУСАКОВСКАЯ ул.
ГОСПИТАЛЬНАЯ ул.
300 м

INTER-TWINING URBAN ECOLOGY

VORST, BELGIUM

Intertwining Urban Ecology is built on a large closed urban block in Brussels. It consists of 59 passive social housing units including houses, apartments and duplexes, a kindergarten, and open spaces. The main objective was to make the area accessible to the public. The volumes are designed as connections to the waiting façades in order to shape the area. The new functions give meaning to the newly formed public spaces that connect the streets. These public spaces have different characters: a wadi in a natural environment, an esplanade, a large urban panoramic staircase, and vegetable gardens. The project pays great attention to sustainability. All buildings are passive and optimize solar gains. Living spaces are not located in the lowest areas, considering the risk of flooding. The ecological public spaces complete a network for bicycles and pedestrians, ensuring permeability and public ownership of the space in the future.

View of Intertwining Urban Ecology from the street and from green public spaces.

ARCHITECTS: BOB361 /// **COMPLETION:** 2020 /// **CLIENT:** SLRB-BGHM, FOYER DU SUD /// **INVESTORS:** PUBLIC /// **LANDSCAPE DESIGN:** ONTWERPBUREAU PAUWELS, LEUVEN /// **CONTEXT:** URBAN /// **GROSS FLOOR AREA:** 7,700 M² /// **NUMBER OF APARTMENTS:** 59 /// **MAIN SUSTAINABLE FEATURES:** THE INTERTWINING OF URBAN TISSUE BY THE CREATION OF ECOLOGICAL AND SLOW PUBLIC SPACES. /// **PHOTOS:** ANDREA ANONI (42 A.R., 44 B.M.) / WWW.ANDREAANONI.COM, ANDRÉ NULLENS / WWW.ARCHIMEDIA.BE

A typical housing unit in the buildings and a view of the open spaces and connecting pathways of the urban block.

A view of the building's bicycle and pedestrian paths and the block from the street at night.

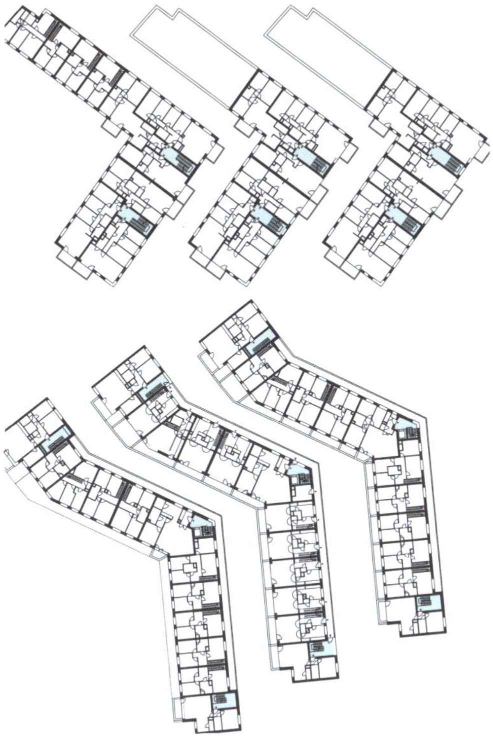

Floor plans and site plan.

A close up view of the building's brick façade and the balconies.

ELEMENT

BERLIN, GERMANY

In an effort to move away from so-called disposable architecture, Studio Loes has realized the Element & Lingot project. The project is characterized by the use of wooden modules and prefabricated concrete elements, which allows for a highly efficient and sustainable construction process. Situated in a densely populated district of Berlin, Element also stands for intelligent densification in the urban environment and serves as an example of future-oriented urban residential architecture. Element consists of five regular floors and one stepped floor, containing 20 apartments, some of which are maisonettes and extend across floors. The reinforced concrete exoskeleton is used to accommodate balconies for the apartments. The timber frame construction façade has been layered between the exoskeleton. Inside, cross-laminated timber walls and ceilings are the load-bearing elements. The building supports cross-ventilation and year-round natural lighting. Furthermore, shared spaces such as lounge areas have been incorporated to stimulate and enable social interaction.

Modules allows for a highly efficient and sustainable construction process.

Close up of the balconies and the wooden façade.

ARCHITECTS: STUDIO LOES /// **COMPLETION:** 2023 /// **INVESTOR:** PRIVATE /// **CONTEXT:** URBAN /// **GROSS FLOOR AREA:** 2, 600 M² /// **NUMBER OF APARTMENTS:** 20 /// **MAIN SUSTAINABLE FEATURES:** WOOD MODULE CONSTRUCTION, PRECAST CONCRETE, EXTENSIVE ROOF GREENING, SOLAR PANELS, HEAT PUMPS, TINY FORESTS /// **PHOTOS:** NATE COOK / WWW.NATECOOKPHOTOGRAPHY.COM

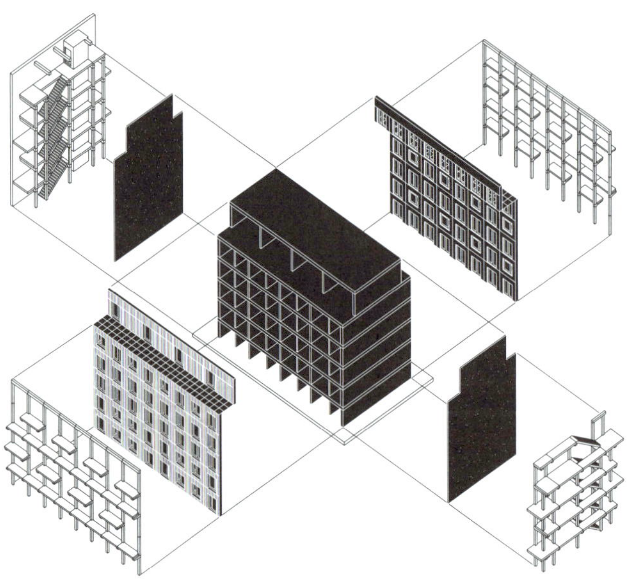

Detailed views of the balconies from varying perspectives.

Axonometry.

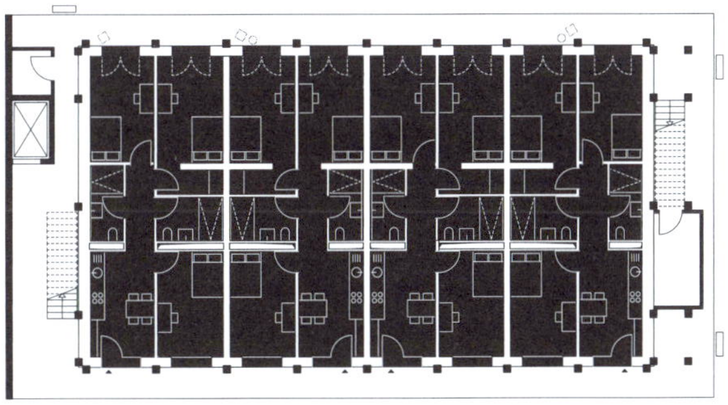

Views of staircases within the building.

Ground floor plan.

PICCOLO HOUSE

MELBOURNE, AUSTRALIA

Piccolo House is a multi-use building that combines residential, retail, hospitality and commercial spaces in one building. Its unique island-like site allows the architecture to be expressed on all sides. Contrasting the restored Victorian building, Piccolo House presents an ordered monolith of off-form concrete, subtly arranged precision cut-outs, and colored reflective glass elements that reference the tones of brickwork in the neighboring lane. The Charles Street and John Street entrances use Corten steel and various other shades of steel to create an interesting entrance area. These are illuminated at night, contributing to a sense of security in the area. Mirrored glass and blackened metal add to the sense of arrival, while the interior of the lobby features a range of autumnal glass tones. Above, four luxurious one- and two-bedroom apartments offer uninterrupted views of Melbourne's skyline, while Piccolo has established its offices on the level above.

Piccolo house rising above the neighboring Victorian building.

ARCHITECTS: WOOD MARSH /// **COMPLETION:** 2020 /// **CLIENT:** PICCOLO /// **INVESTOR:** PRIVATE /// **ORIGINAL BUILDING:** HERITAGE VICTORIAN BUILDING, 1880 /// **CONTEXT:** URBAN /// **GROSS FLOOR AREA:** 4,000 M^2 /// **NUMBER OF APARTMENTS:** 24 /// **PHOTOS:** TREVOR MEIN / WWW.TREVORMEIN.COM

Elevation.

A view of the building from a street level, show-casing the access from Charles street.

Views of the roof terrace and from within an apartment and the office space.

The entrance area, kept in steel shades, illuminated at night.

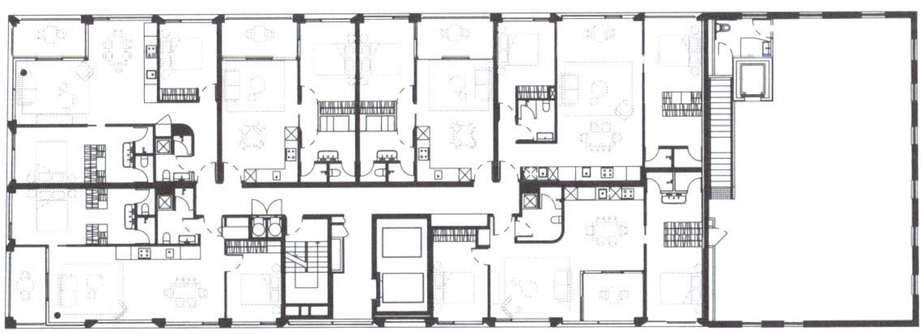

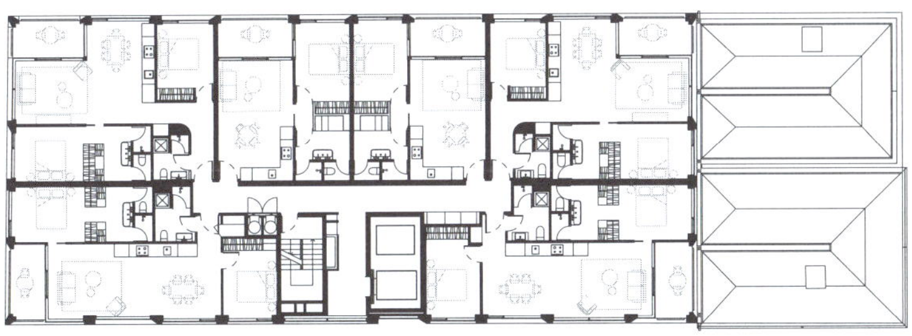

Typical floor plans.

PRIVATE PROPERTY
Unauthorised vehicles will
be towed away at owners expense

The south-west view of the building showcasing its impressive silhouette and the different façade designs with the balcony variations.

ARCHITECTS: STEFAN FORSTER GMBH /// **COMPLETION:** 2024 /// **CLIENT:** HANOVA WOHNEN GMBH /// **INVESTOR:** PUBLIC /// **LANDSCAPE DESIGN:** ACKERMANN, BÜRO FÜR GARTEN UND LANDSCHAFTSARCHITEKTUR BDLA /// **CONTEXT:** URBAN /// **GROSS FLOOR AREA:** 5,500 M² /// **NUMBER OF APARTMENTS:** 50 /// **MAIN SUSTAINABLE FEATURES:** USE OF ENVIRONMENTALLY FRIENDLY MATERIALS /// **PHOTOS:** LISA FARKAS / WWW.LISAFARKAS.DE

A closer view of the building's façade from two perspectives.

WOHNEN AM KRONS-BERG SÜD

HANNOVER, GERMANY

This residential and commercial building is located at the southern entrance to the Kronsberg Süd development area. With its triangular floor plan and sloping top, it forms a public square to the south and a communal courtyard for the residents to the north. The inwardly angled façade and the contrasting loggia parapets made of folded steel take up the triangular motif. Steel forms a deliberate contrast to the warm red clinker brick façade, which gives the building a three-dimensional structure with rolled layers and deep window reveals and refers to local building traditions. All apartments have a private outdoor area in the form of loggias or terraces. Above the building's base, which is used by three commercial units, there are two-to-four-room apartments with different floor plans and layouts. The mixed use in the building is complemented by the combination of subsidized and privately financed apartments.

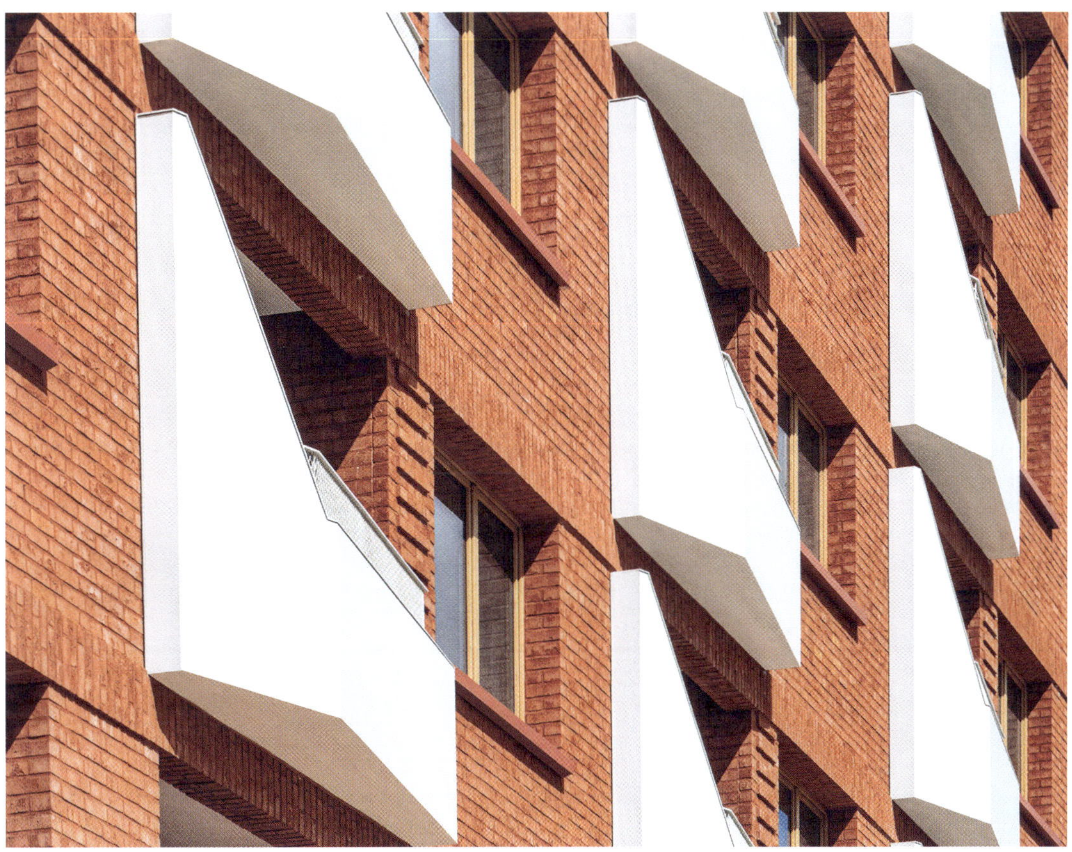

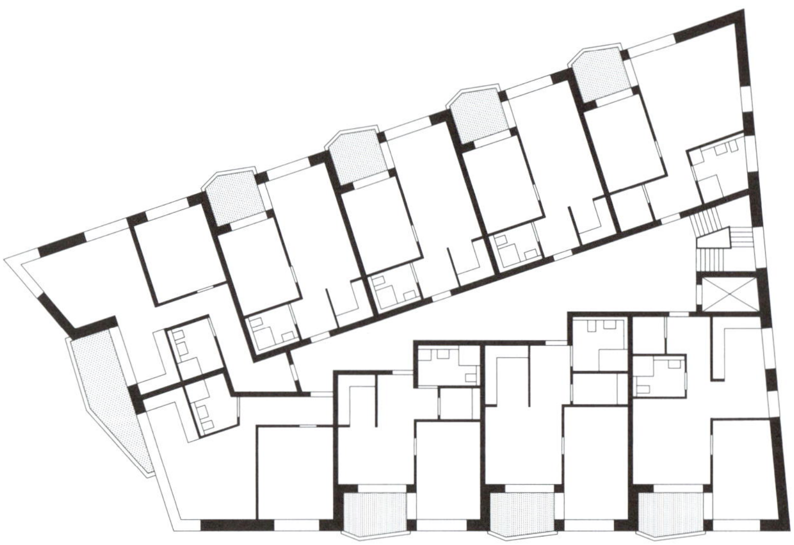

A close up view of the balconies on the building#s front side.

Floor plan.

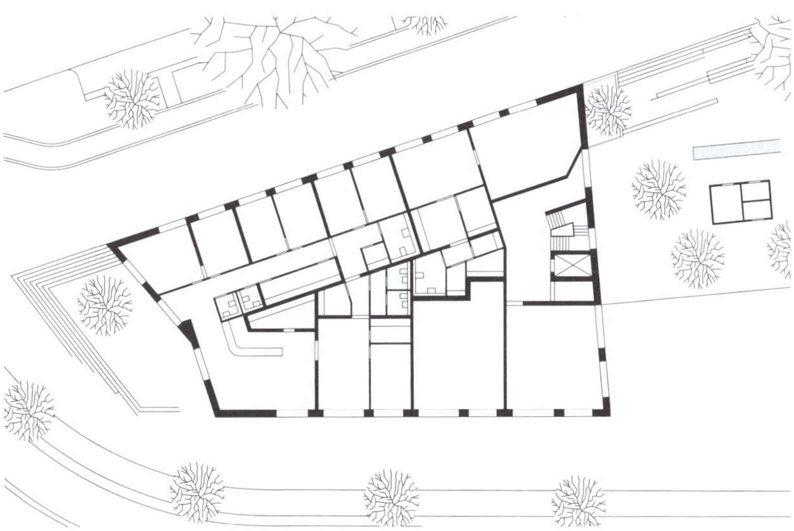

The staircases inside the building.

Ground floor plan.

EOS

VARNA, BULGARIA

Built in the Chaika neighborhood of Varna, this building was named after the Greek goddess of dawn. In the mythology, the goddess brings light to humans and gods alike. The trigonal shape of the plot sets a unique frame, where the designers constructed a building that combines the pragmatic need for rooms to live and work in whilst still capturing and framing the stunning landscape of the surroundings. By post-tensioning after concreting, the balconies were significantly expanded in area, triumphing over the natural forces. Precisely placed steel cables counteract the projected loads. These technologies are means to break the sky's limit, creating a feeling of boundlessness in the structure with the highest level rising up to eight meters above the ground. The high-rising structure shapes the silhouette of the neighborhood, while acting as a kind of guardian for the bay over which it is watching. Three levels of the building are used as office spaces, whereas the higher levels are reserved for vacation apartments.

View of the building from the main boulevard.

ARCHITECTS: STARH /// **COMPLETION:** 2022 /// **CLIENT:** RID /// **CONTEXT:** URBAN /// **GROSS FLOOR AREA:** 6,170 M² ///
NUMBER OF APARTMENTS: 39 /// **PHOTOS:** DIAN STANCHEV / WWW.DIANSTANCHEV.COM

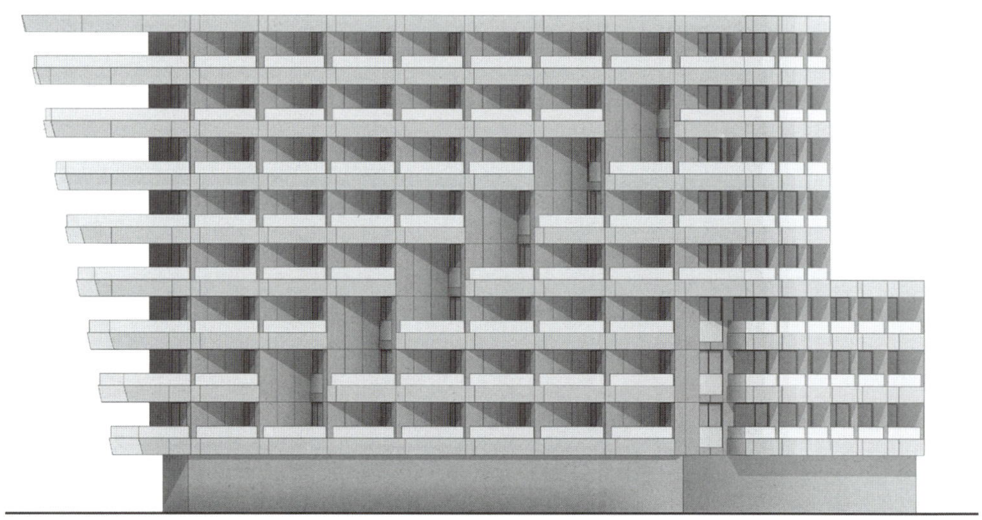

EOS offers an uncommon appearance from different points.

Elevation.

Terraces on the building's eastern corner project outward.

Cross section.

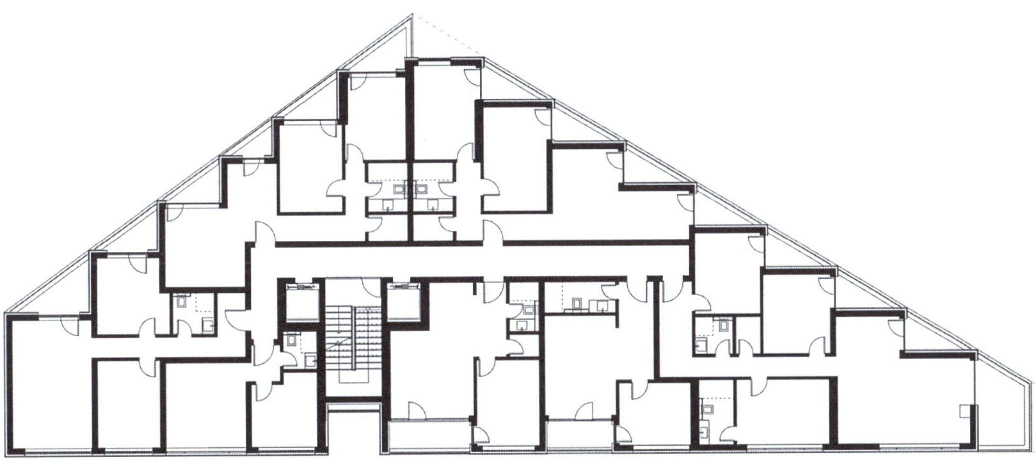

ON THE HILLSIDES

CHAVILLE, FRANCE

This innovative project, comprising 20 social housing units and a 26-place boarding house, stands out for its harmonious integration into the urban fabric and its respect for the environment. Optimum use has been made of the complex topography of the site to create generous green spaces and unobstructed views, while blending gently into the surrounding neighborhood. Each home benefits from abundant natural light, private outdoor spaces and landscaping designed to enhance the well-being. Architectural flexibility responds to the specific challenges of the site, in particular the slope and shape of the land, creating functional spaces that encourage both social interaction and the preservation of privacy.

The choice of sustainable materials such as wood and brick underlines the ecological and aesthetic commitment, helping to integrate the project into its environment. This project is the result of close collaboration between the architects, the municipality, and social partners, illustrating the importance of a holistic approach to urban housing.

Main façade facing the street, linking the existing buildings and the narrow alleyway.

ARCHITECTS: AUP ARCHITECTES /// **COMPLETION:** 2023 /// **CLIENT:** HAUTS DE SEINE HABITAT /// **INVESTOR:** PUBLIC /// **LANDSCAPE DESIGN:** D&H PAYSAGES /// **CONTEXT:** URBAN /// **GROSS FLOOR AREA:** 2,291 M² /// **NUMBER OF APARTMENTS:** 46 /// **MAIN SUSTAINABLE FEATURES:** FRENCH LABELS E2C1, HQE 7* /// **PHOTOS:** 11H45 / WWW.11H45.COM

Street elevation.

The rear façade, overlooking the hills, undulates and adapts to the geometry of the land.

Cross section.

Frontal view of the urban brick façade, where the play of volumes creates a dialogue with the existing building.

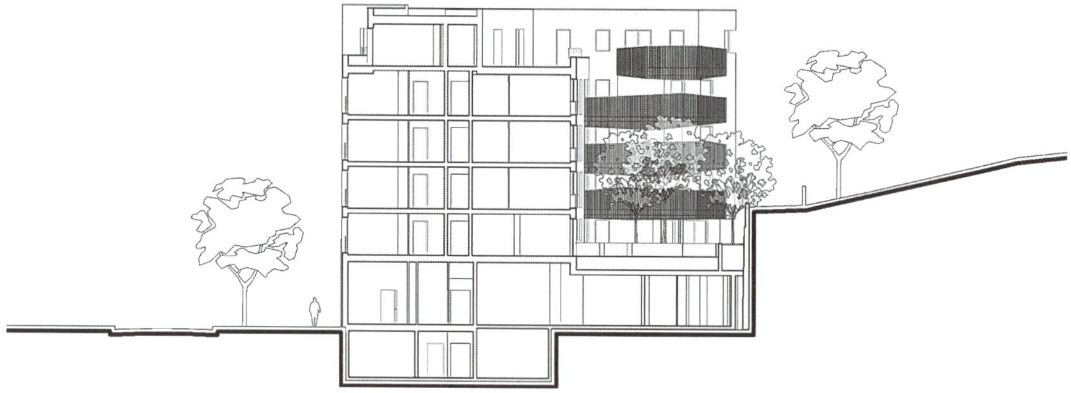

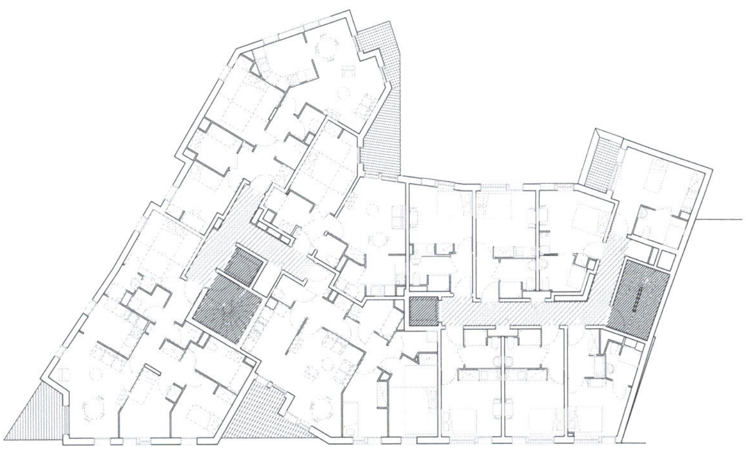

A courtyard at the heart of the block where the wooden façades expand. Facing the courtyard, the wooden louvred balconies play with light and privacy.

Ground floor plan.

WOHNHAUS H4

DÜSSELDORF, GERMANY

The H4 residential building is located in Düsseldorf-Gerresheim, directly adjacent to the listed Protestant community center. The new building forms the transition between the closed Gründerzeit style development to the free-standing residential development. This transition is achieved with a differentiated height staggering and sloping roof surfaces to the respective connection heights. Designed as two- and three-bedroom units, a total of nine apartments with individually divided floor plans have been created here, heated by an air heat pump located on the roof. Ventilation takes the form of controlled living space ventilation to limit transmission heat loss. The exhaust air is discharged via the adjoining rooms, while the supply air flows in via window rebate ventilators. The building design takes into account optional, time-delayed demolition versions. The diaconal housing group, which is located in a large unit on the ground floor and basement, can be converted into two maisonette apartments.

The main façade along the street, canopy of the entrance, and a corner view showing the appartments in the roof.

ARCHITECTS: PIER7 ARCHITEKTEN /// **COMPLETION:** 2024 /// **CLIENT:** EVANGELISCHE KIRCHENGEMEINDE DÜSSELDORF GERRESHEIM /// **CONTEXT:** URBAN /// **GROSS FLOOR AREA:** 1,645 M^2 /// **NUMBER OF APARTMENTS:** 9 /// **MAIN SUSTAINABLE FEATURES:** AIR HEAT PUMP, CONTROLLED LIVING SPACE VENTILATION /// **PHOTOS:** MICHAEL REISCH / WWW.MICHAELREISCH.NET

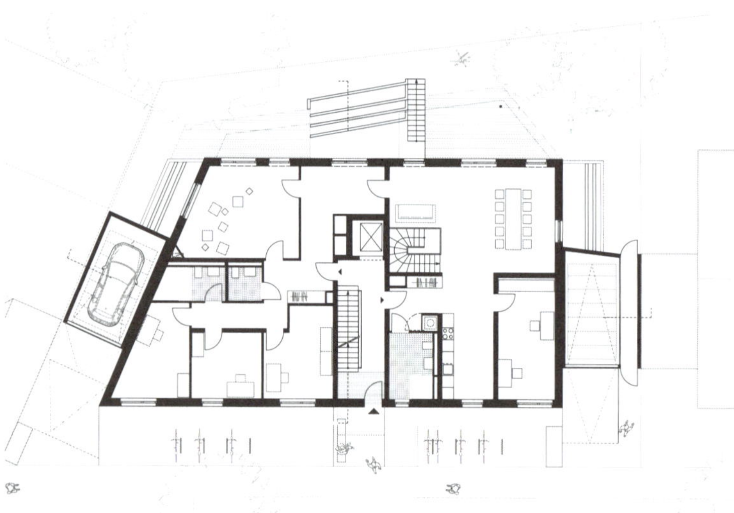

View to the balcony and the staircase in one of the two maisonette apartments.

Site plan with ground floor plan.

Rear view of the building and a close up of the windows in the living room.

THREE SISTERS

HAMBURG, GERMANY

The Three Sisters-building was built on the 1,826-square-meter plot called Reeseberg 115 in the south of Hamburg. Originally, the property housed a supermarket for the residential area that developed in the 1960s. Today, three diamond-shaped buildings, one five-story and two four-story, stand on the site. The central idea in developing these buildings was to create a cost- and space-efficient apartment complex. The buildings were constructed of reinforced concrete and sand-lime brick, while the roofs were extensively planted to improve the local microclimate. The placement of the windows varies slightly from floor to floor making the design seem less static. Wrapped in a façade of champagne-colored aluminum, the buildings house 37 apartments and a separately accessible kindergarten which offers access to extensive green areas for the children to play in. The building also includes a shared bike room in the basement and a playground, further contributing to a cohesive neighborhood community.

The buildings have a coherent appearance from any perspective due to their diamond shape.

A street level view of the building showing the shifting window placements, balconies, and an entrance area.

ARCHITECTS: RENNER HAINKE WIRTH ZIRN ARCHITEKTEN /// **COMPLETION:** 2023 /// **CLIENT:** BAUVEREIN REIHERSTIEG EG /// **INVESTOR:** PRIVATE /// **CONTEXT:** SUBURBAN /// **GROSS FLOOR AREA:** 1,826 M² /// **NUMBER OF APARTMENTS:** 37 /// **MAIN SUSTAINABLE FEATURES:** GERMAN KFW EFFIZIENZHAUS, GREEN ROOF, HEAT PUMPS, THREE PANES OF THERMAL INSULATION GLASS /// **PHOTOS:** STEPHANIE BRINKKÖTTER / WWW.BRINKKOETTER.COM

The entrance area of the apartment buildings is distinct from the rest of the buildings. Stone is used in contrast to the champagne colored aluminum façade and protects against soiling.

Ground floor plan.

The entrance area to each building is illuminated at night. The kindergarten located in one of the buildings with a separate entrance.

MAXACHT

STUTTGART, GERMANY

MaxAcht is a multi-generational housing community for people of all age groups. Eleven apartments of varying sizes have been built for households of one to four people, including two subsidized condominiums and two inclusive apartments. The aim was to create a lively community where people could live, celebrate and, if necessary, care for each other. The building is designed so that the layouts can be adapted over time. Large flats can be partitioned off for home care. In general, MaxAcht is designed to be barrier-free. It deliberately stands out from the heart of the neighborhood, demonstrating sustainable, ecological, and communal living with high quality architecture. A building shell made of pre-grayed spruce formwork gives the building an urban character. The timber construction remains visible in all apartments, creating a pleasant and healthy living environment. Public access is kept to a minimum, allowing maximum use of the site for the benefit of the living space, thus achieving maximum living quality. MaxAcht allows easy removal and reuse of material.

MaxAcht terrace and balconies on pedestrian corner.

ARCHITECTS: ARCHITEKTURAGENTUR /// **COMPLETION:** 2019 /// **CLIENT:** WEG MAXACHT GBR /// **INVESTOR:** PRIVATE/// **CONTEXT:** URBAN /// **GROSS FLOOR AREA:** 1, 725 M² /// **NUMBER OF APARTMENTS:** 11 /// **MAIN SUSTAINABLE FEATURES:** CEM III REINFORCED CONCRETE, REMOVABLE AND REUSABLE LOAD BEARING STRUCTURE AND FAÇADE MADE OF WOOD, GREEN ROOF /// **PHOTOS:** JÜRGEN POLLAK / WWW.JUERGENPOLLAK.DE

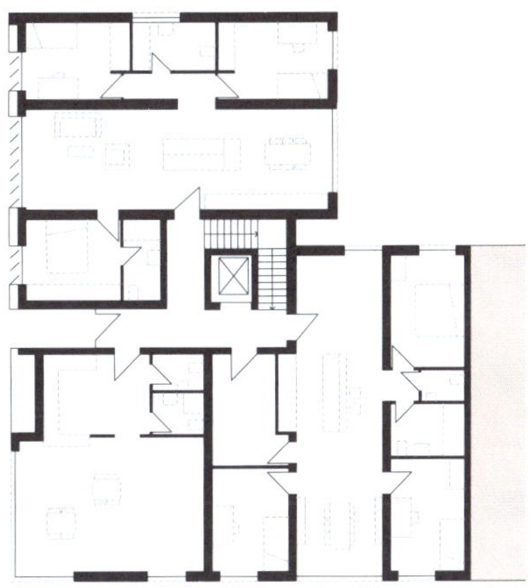

Ground Floor Plan.

Angled view of MaxAcht.

Typical interiors of the eleven apartments.

View of the building's wooden façade and a bathroom.

Typical floor plan.

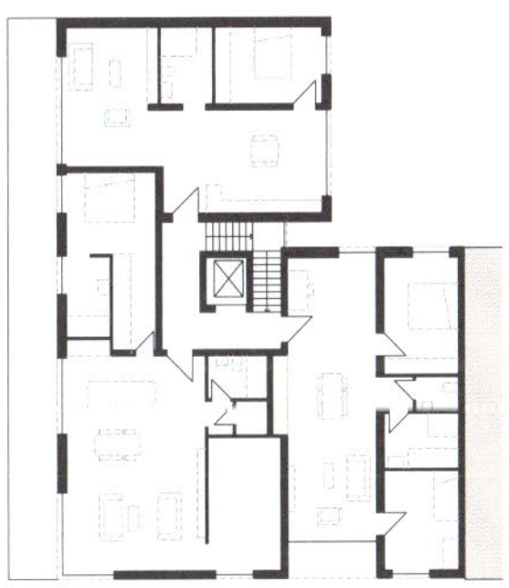

UPPER HOUSE

BRISBANE, AUSTRALIA

Upper House in South Brisbane, Australia, is an expressive multi-residential tower that houses 188 apartments over 33 stories. The architectural form is inspired by native flora and its design signals a new era in urban multi-residential development, where connection to nature, resident wellbeing, and environmental sustainability align. Upper House punctuates the city skyline with a timber pergola and tropical rooftop oasis that is designed to make the most of Brisbane's subtropical climate. Progressive resident amenities with a focus on interpersonal connection and wellbeing are housed on the top two levels, while at the street level, a five-story indigenous artwork contributes an important message of Australia's rich ancient history to the neighborhood. The façade is inspired by the roots of the Moreton Bay Fig, symbolizing a journey from the ancient origins to the future. These facilities for wellbeing are also social hubs and seek to reconnect the urban population and create a sense of belonging within a community.

Five-story indigenous artwork by Judy Watson. In total, over 108 original art pieces were commissioned for the project.

ARCHITECTS: KOICHI TAKADA ARCHITECTS /// **COMPLETION:** 2023 /// **CLIENT:** ARIA PROPERTY GROUP /// **INVESTOR:** PRIVATE /// **CONTEXT:** CITY CENTER /// **GROSS FLOOR AREA:** 19,700 M² /// **NUMBER OF APARTMENTS:** 188 /// **MAIN SUSTAINABLE FEATURES:** 5-STAR GREEN STAR RATED, EXTENSIVE LANDSCAPING (3,544 PLANTS), RAINWATER COLLECTION, ROOFTOP SOLAR , ELECTRIC VEHICLE RESIDENT CAR SHARE, 100% RENEWABLE ENERGY PROCURED /// **PHOTOS:** SCOTT BURROWS / WWW.SCOTTBURROWSPHOTOGRAPHER.COM (85B., 86A., 87B.) / MARK NILON (84, 86B., 87A.)

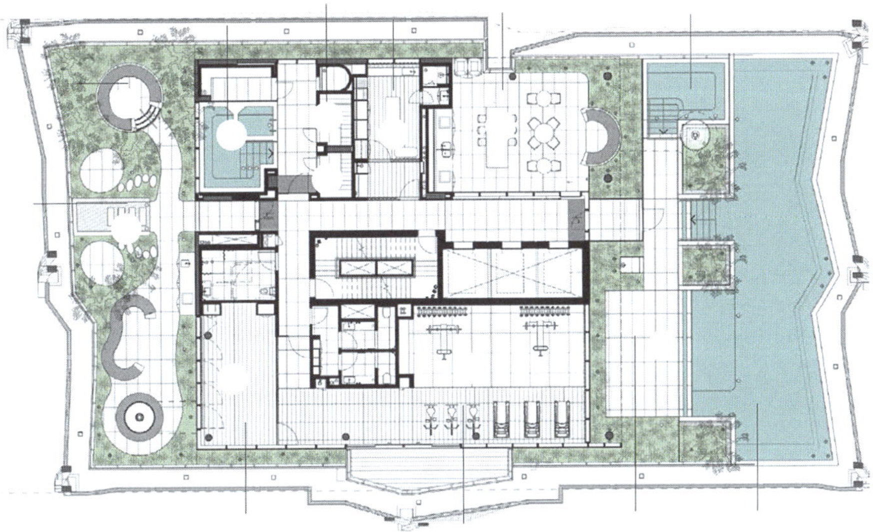

Floor Plan.

The timber pergola is an organic contrast to the rectilinear city forms beyond.

The Lobby operates as a public gallery to show-case emerging artists, extending the reach of Fish Lane Arts Precinct. A Wellness Club and social facilities span the top-two levels, encouraging residents to interact and foster community.

Two timber nests are cosseting vantages from which to enjoy the Brisbane riverside and expansive views. The façade artwork by Judy Watson was conceived as a way to contribute a valuable narrative to the urban landscape.

DOCKLEY APART-MENTS

LONDON, UNITED KINGDOM

On a former industrial site, 111 apartments have been created, as well as industrial spaces for artisans. The railway arches are home to a unique community of merchants, artisan food producers, and wholesalers who open their doors to the public every Saturday. A vibrant urban block has been created with a variety of ground floor activities that complement those within the railway arches. The building offers over 35 unique apartment layouts while achieving a benchmark of 38 percent affordable housing. The apartments are arranged around a central courtyard that acts as a landscaped, communal space for relaxation and play. Residents access their apartments via wide galleries overlooking the courtyard, which extend to communal roof terraces on the upper floors. These are designed as places to sit, share and enjoy views of the city skyline. All apartments are dual aspect. Each apartment has a generous outdoor balcony that balances the need for summer shade with access to daylight and outdoor dining and entertaining.

New building in the existing context and view from one apartment terrace to others.

Building form driven by existing trees where stepped form creates generous balconies and terraces .

ARCHITECTS: STUDIO WOODROFFE PAPA, POGGI ARCHITECTURE /// **COMPLETION:** 2022 /// **CLIENT:** MATCHING GREEN /// **INVESTOR:** PRIVATE /// **CONTEXT:** URBAN /// **GROSS FLOOR AREA:** 10,678 M^2 /// **NUMBER OF APARTMENTS:** 111 /// **MAIN SUSTAINABLE FEATURES:** DUAL-ASPECT, CROSS VENTILATED, GENEROUS PRIVATE OUTDOOR AMENITY. /// **PHOTOS:** TIM CROCKER / WWW.TIMCROCKER.CO.UK

A new laneway creates a social space, as well as a place for artisans to work. Sheltered play spaces in the courtyard.

Undulating profiles providing sheltered ways to the apartments. The upper levels extend onto generous outdoor terraces.

The corners broaden the variety of apartment types.

Ground floor plan with isometric view.

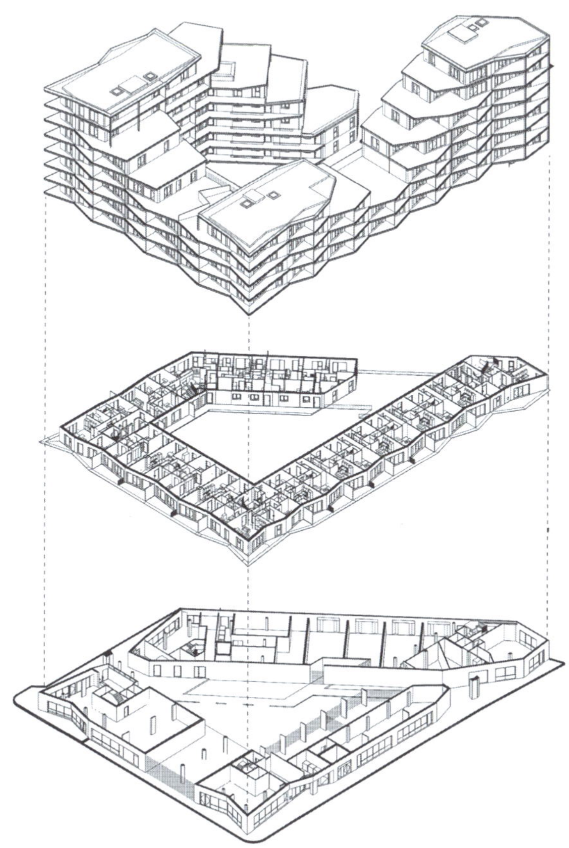

GRANVILLE 1500

LOS ANGELES, CA, USA

The new mixed-use development provides affordable housing alongside academic, commercial and public functions. Located just ten minutes from UCLA's main campus, the project blurs the boundaries and embeds academic life into the urban context in which it is situated. On this site of a former car dealership, Granville1500 aims to reinforce a sense of place while adding to the vitality and energy of this fast-paced, emerging neighborhood. The design breaks with the bulk typical of most mega-block housing projects by showcasing the residential and pedestrian experience at street level. Instead of a massive, imposing façade, three discrete wedge-shaped volumes appear along Santa Monica Boulevard. A series of large pyramidal carve-outs, shaped like inverted prisms, have been placed at the corners of the complex. Granville1500 reinforces the architects' commitment to creating spaces for civic engagement in urban development. The project embraces the culture and ecology of its surroundings and provides opportunities to engage with the neighborhood.

Main view from Santa Monica Boulevard.

ARCHITECTS: LORCAN O'HERLIHY ARCHITECTS /// **COMPLETION:** 2021 /// **CLIENT:** STARTED BY CIM GROUP, TAKEN OVER BY UCLA /// **INVESTOR:** PRIVATE /// **LANDSCAPE DESIGN:** LA GROUP INC. /// **INTERIOR DESIGN:** ARIEL FOX DESIGN /// **CONTEXT:** CITY CENTER /// **GROSS FLOOR AREA:** 29,012 M^2 /// **NUMBER OF APARTMENTS:** 153 /// **MAIN SUSTAINABLE FEATURES:** HIGH ALBEDO COOL ROOF, NATURALLY VENTILATED CORRIDORS, DROUGHT TOLERANT LANDSCAPING, STORMWATER MANAGEMENT PLAN, SOLAR PANELS /// **PHOTOS:** PAUL VU / WWW.HEREANDNOWAGENCY.COM

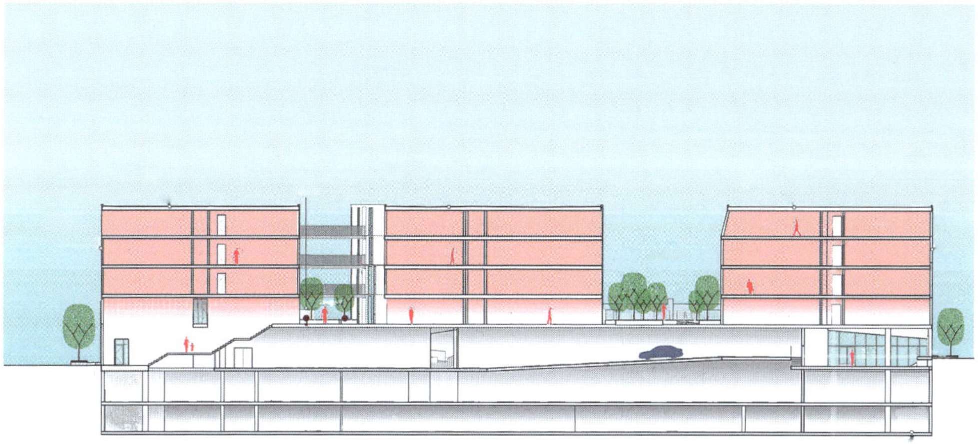

Cross section.

The volumes, connected by bridges, and the communal courtyard.

Various views of the buildings' interior spaces.

Street level views of the pyramidal carve-outs at the buildings' corners widening the sidewalk.

Ground floor plan.

A perspective of the building from inside the courtyard.

ARCHITECTS: POLE ARCHITEKCI /// **COMPLETION:** 2022 /// **CLIENT:** MOTA-ENGIL /// **INVESTOR:** COOPERATIVE /// **CONTEXT:** URBAN /// **GROSS FLOOR AREA:** 1,400 M^2 /// **NUMBER OF APARTMENTS:** 12 /// **PHOTOS:** FILIP BRAMORSKI / WWW.FILIPBRAMORSKI.PL

LA SCALA

WARSAW, POLAND

Situated on a narrow site in Warsaw, La Scala's use of recessed balconies and entrances, which move the external walls away from neighboring buildings, efficiently creates more space and thus a more comfortable living environment for the occupants. The shape of the building was largely determined by functionality, but also reflects other considerations, such as the desire to provide occupants with privacy and plenty of sunlight in a densely built urban environment. This is visible in the unique arrangement of the windows. The building's square window openings increase in size towards the building's entrance and terraces and decrease in size towards neighboring buildings. The light tiles are also deliberately designed to reflect sunlight efficiently. La Scala's design fulfills the needs of its occupants while keeping the environment in mind. The rear of the building opens onto a green courtyard with numerous trees providing shade and cooling.

Street level views of La Scala, highlighting its unique window arrangement and recessed entrances.

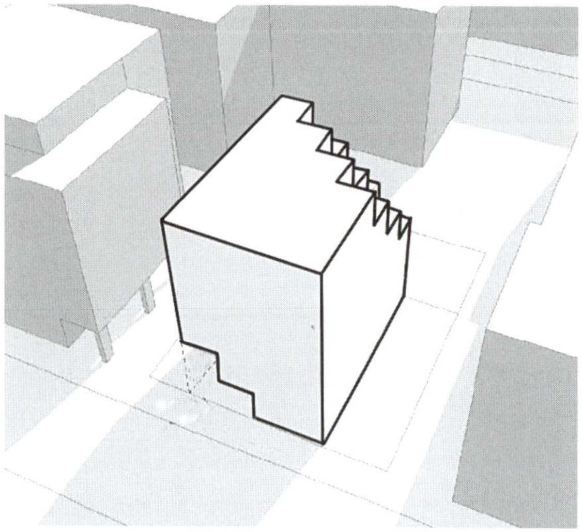

A rear view of the building and a close up view of the building's light-colored tiles as well as the varying window sizes.

Building perspective.

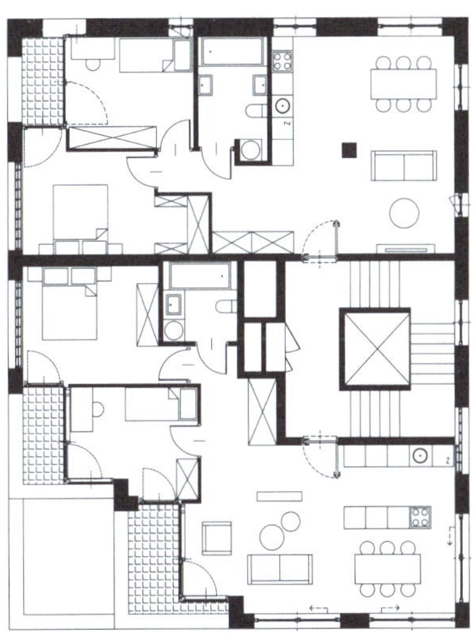

Another perspective of the recessed balcony design as well as a diagonal view of the building.

Typical floor plan.

NIRAN APARTMENTS

NAKHON PATHOM, THAILAND

Located near Nakhon Pathom's Silpakorn University, Niran Apartments looks like a typical Nakhon Pathom block of flats. Its exterior has been deliberately designed to be unremarkable and blend in with its surroundings. This is in stark contrast to its interior, which has been carefully designed to be a highly unique space. The atrium is designed as a public meeting place. It was also important to accommodate as many units as possible, while leaving room for some saleable space. The solution was to create three different types of spaces at varying depths. The public and retail spaces are located on the first two floors, while the smallest units are located on the third floor, the lowest of the residential floors, creating more space in the atrium, with the larger apartments on the higher floors. This creates the iconic stupa-inspired A-shape. The steel bar façade is woven into 22 unique butterfly-shaped panels. It divides spaces and creates more privacy, while deliberately leaving open spaces to facilitate connection and interaction.

The unique stupa inspired A-shaped atrium inside Niran Apartments.

ARCHITECTS: RAD STUDIOS /// **COMPLETION:** 2020 /// **CLIENT:** DURONGJIRAKAN RUNGRUEANG COMPANY LIMITED /// **INVESTOR:** PRIVATE /// **CONTEXT:** URBAN /// **GROSS FLOOR AREA:** 7,000 M^2 /// **NUMBER OF APARTMENTS:** 72 /// **MAIN SUSTAINABLE FEATURES:** SOLAR PANELS, STACK VENTILATION SYSTEM /// **PHOTOS:** CHALERMWAT WONGCHOMPOO / WWW.SOFOGRAPHY.COM

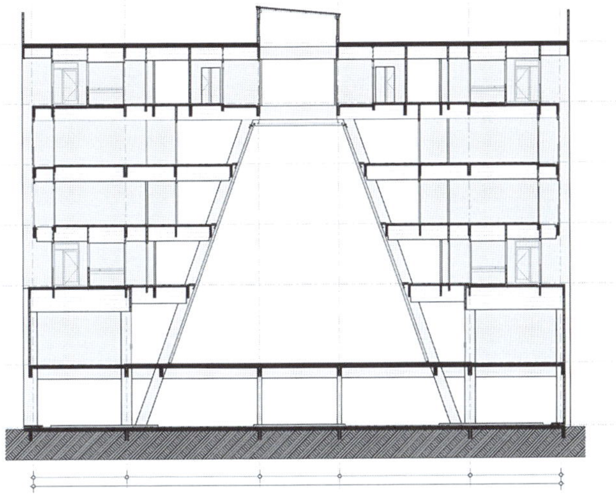

Cross Section.

One of the openings in the façade that lead to the atrium.

The exterior of the Niran Apartments, a close-up of its steel-rod façade butterfly panels, and a typical apartment.

CITÉ ANGUS PHASE II

MONTREAL, CANADA

Cité Angus Phase II is located at the head of the central block of Technopôle Angus in Montreal. As a six-story residential building with commercial spaces on the ground floor, the project's sustainable design includes quality-of-life considerations for occupants and visitors. The building takes the form of two volumes articulated around an interior courtyard animated by a spiral staircase that connects all floors via a system of external walkways. This space serves to provoke social interactions and encourages active transport by allowing residents to access their apartments, overlooking the courtyard. The walkway system maximizes natural light and ventilation, provides high quality views and creates a variety of residential typologies. The private outdoor spaces vary in size, location and relationship to their local context. The project thus responds to the diverse and evolving needs of today's families and fits into an integrated vision of urban development.

Close view of the front façade, showcasing the use of a metal mesh as a second skin, and full view of the front façade.

ARCHITECTS: ÆDIFICA /// **COMPLETION:** 2023 /// **CLIENT:** SOCIÉTÉ DE DÉVELOPPEMENT ANGUS /// **INVESTOR:** PRIVATE /// **LANDSCAPE DESIGN:** NIP PAYSAGE /// **CONTEXT:** URBAN /// **GROSS FLOOR AREA:** 10,590 M^2 /// **NUMBER OF APARTMENTS:** 88 /// **MAIN SUSTAINABLE FEATURES:** LEED ND CERTIFICATION FOR THE SITE, LEED BD+C MULTIFAMILY MIDRISE FOR THE BUILDING /// **PHOTOS:** DAVID BOYER / WWW.DAVID-BOYER.CA

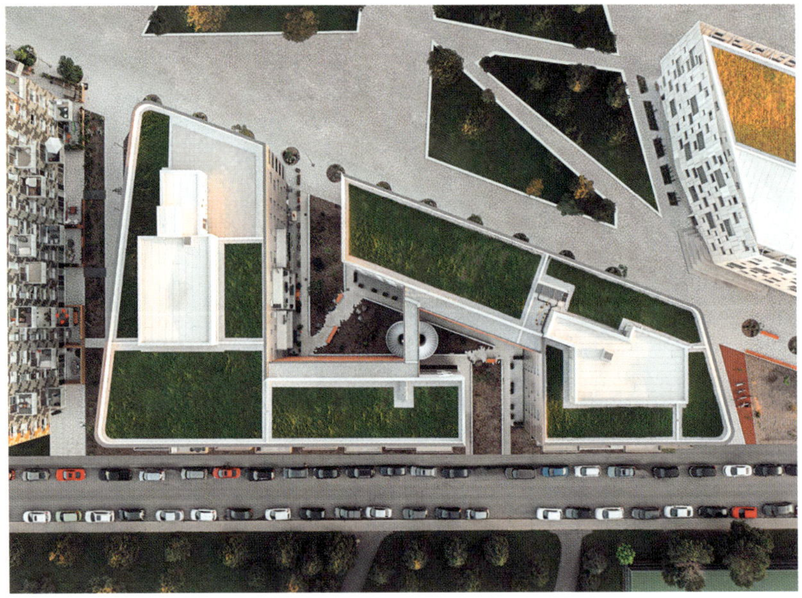

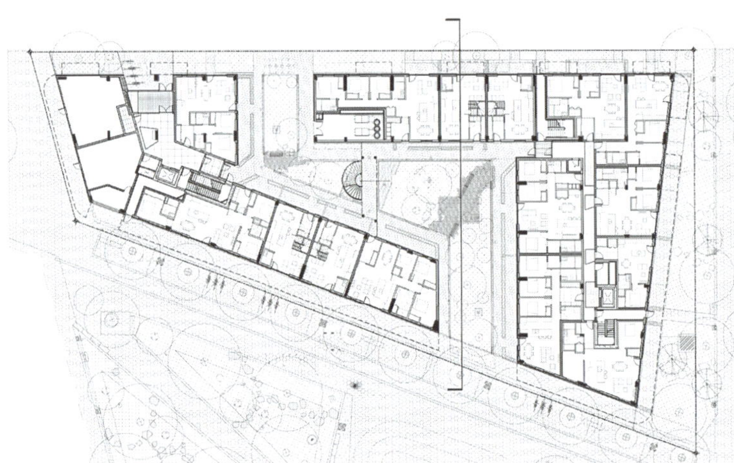

A general view of the building from above, and a close-up of the metal mesh that acts as a second layer to the building's surface.

Ground floor plan.

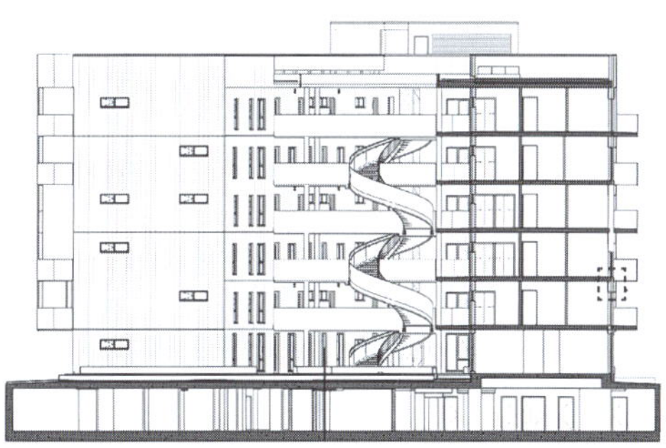

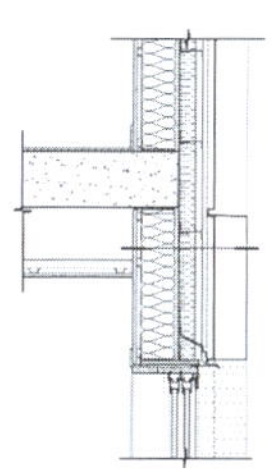

The helical staircase and exterior walkways in the interior courtyard.

Cross section.

JAURÈS PETIT HOUSINGS

PARIS, FRANCE

Located on a plot of land in the heart of the block occupied by a car park, the project comprises 74 social housing units and proposes to involve both the developers and the residents in the steps of a rational, visionary and sustainable approach to urban transformation. The building has a wooden structure. The new wood-frame operation of Petit Jaurès Housings offers a structure adapted to the specifications of the client. The fenestration and balconies allow for ample sunlight and ventilation for each apartment. The gardens in the center of the site are defined by spatial continuity. A number of little garden areas are scattered around the site and along the path. The pedestrian paths run alongside the buildings throughout the site. The architects' intention was to leave the garden as spontaneous and varied as possible after planting. Trees punctuate the wide open spaces, with rather light foliage, chosen for their flowering as an annual sign of renewal.

General view at night.

ARCHITECTS: ARCHI5 /// **COMPLETION:** 2021 /// **CLIENT:** PARIS HABITAT /// **INVESTOR:** PUBLIC /// **LANDSCAPE DESIGN:** ATELIER ROBERTA /// **CONTEXT:** URBAN /// **GROSS FLOOR AREA:** 4,874 M² /// **NUMBER OF APARTMENTS:** 74 /// **MAIN SUSTAINABLE FEATURES:** WOODEN BUILDING STRUCTURE, WOOD CLADDING, ENERGY CONSUMPTION < 50 KWH/M²/YEAR /// **PHOTOS:** SERGIO GRAZIA / WWW.SERGIOGRAZIA.FR

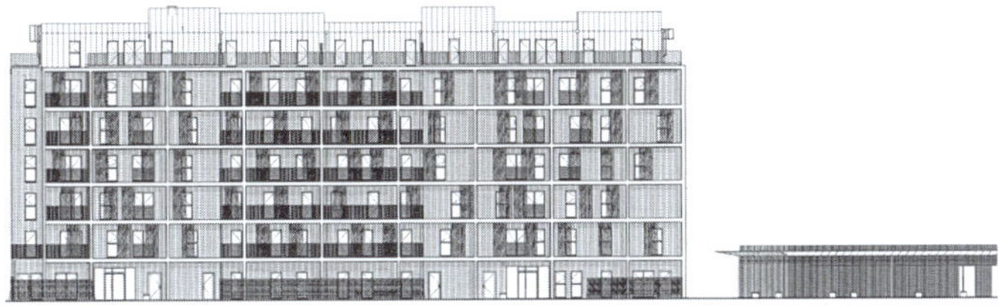

East façade plan.

Entrance area and the accessing pathway.

Main front view of the building, the pathway, and a typical apartment room.

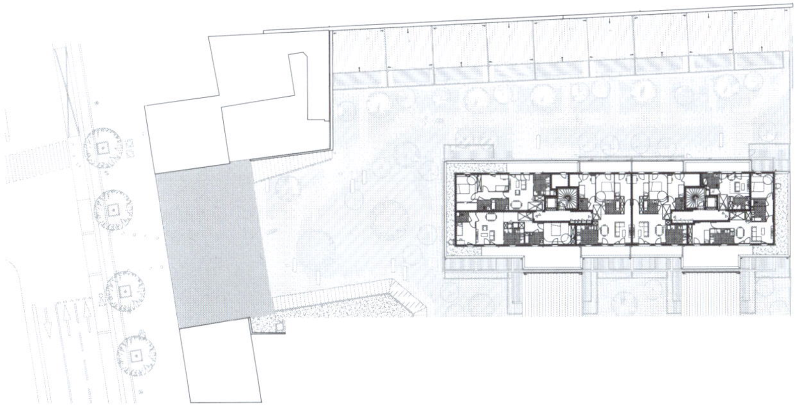

Detailed balcony and façade close ups.

Typical floor plan.

The future park offers unobstructed views of the Boulevard Colonel Fabien. The metal fencing was transformed into a work of art by street artist C215.

ARCHITECTS: BOB361 /// **COMPLETION:** 2023 /// **CLIENT:** CRÉDIT AGRICOLE IMMOBILIER CORPORATE ET PROMOTION /// **INVESTOR:** PRIVATE /// **CONTEXT:** URBAN /// **GROSS FLOOR AREA:** 4,518.97 M^2 /// **NUMBER OF APARTMENTS:** 54 /// **MAIN SUSTAINABLE FEATURES:** RESILIENCE TO FLOOD RISKS /// **PHOTOS:** FABRICE BESSON / WWW.FABRICEBESSON.COM (116, 117. B.L.,118-119), LUC BOEGLY / WWW.BOEGLY-GRAZIA.COM (117.A.R., 120 A.L., 121 A.R.)

IVRY 30 TOWERS

IVRY-SUR-SEINE, FRANCE

IVRY 30 is situated in the immediate vicinity of a large park, in the ZAC Ivry Confluences on the banks of the Seine, just outside Paris. The two towers comprise 32 homes for residential sale and 22 social housing units, underground parking, bicycle storage, and a commercial co-working space. Its location near the Seine and the risk of flooding caused by climate change, induce strict regulations that benefit the covered entrance halls - generous large empty spaces. Designed as convivial spaces, they encourage interaction between social housing residents, homeowners and commercial users. Together with the gardens, they form a unique and collective whole. The architecture of the project enters into a dialogue with the EDF monuments designed by Atelier de Montrouge in the 1960s, and together they form a façade of the park. The entire project is made of white concrete, a noble and luminous material. The double-height terraces and the compact design allow for large living spaces with multidirectional views.

The double-height terraces were sculpted to perfectly frame the view. View of the artwork on the metal fence from the ground floor, where communal spaces were created.

Bou

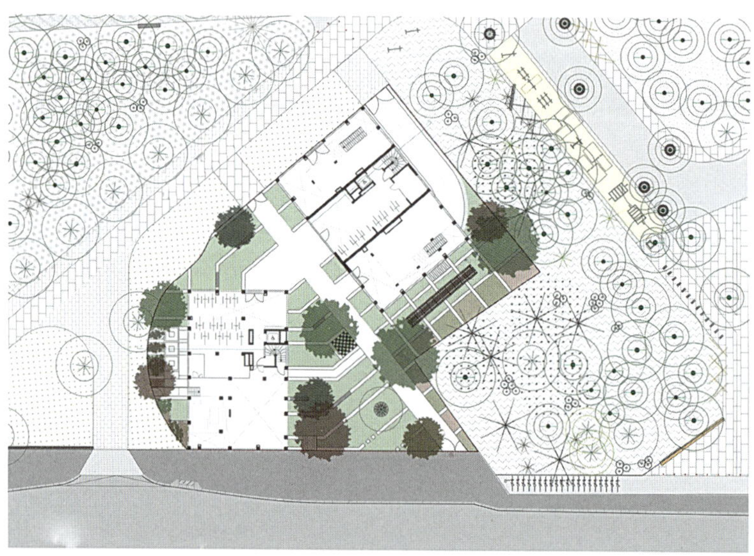

View from the interior showing the gaps in between the balconies to allow the observation of the city life.

Site plan.

Views of the façade inspired by the EDF Tower.

Section.

SAKURAGI-CHO RESIDENCE

YOKOHAMA, JAPAN

This project redefines exterior wall tiling and balcony construction in large Japanese apartment blocks. Typically, these blocks feature continuous balconies for emergency evacuation and air-conditioning unit space. While a retail store occupies the ground level, the façade's uniformly aligned openings are masked to blur residential and commercial uses. In Yokohama, there's a preference for classical aesthetics, echoed in a nationwide trend since the 1980s to use exterior tiling to enhance apartment marketability. Here, instead of traditional tiling prone to weathering, a 15mm grooved pattern mimics brickwork, fitting the classical local style. Balconies are given a warm wood finish using a texturing trowel that simulates wood grain. The design's grid-like structure with grooved walls and woodgrain paint adds both aesthetic and asset value, blending modern urban architecture with historical influences.

Front façade view of the building showcasing the grid-like structure.

Street level view of the building contextualizing the surrounding area.

ARCHITECTS: KEY OPERATION INC./ARCHITECTS /// **COMPLETION:** 2021 /// **CLIENT:** TOHSHIN PARTNERS /// **INVESTOR:** PRIVATE /// **CONTEXT:** CITY CENTER /// **GROSS FLOOR AREA:** 4,129 M² /// **NUMBER OF APARTMENTS:** 109 /// **PHOTOS:** TOSHIYUKI YANO

Views of the entrance area from inside and outside of the building.

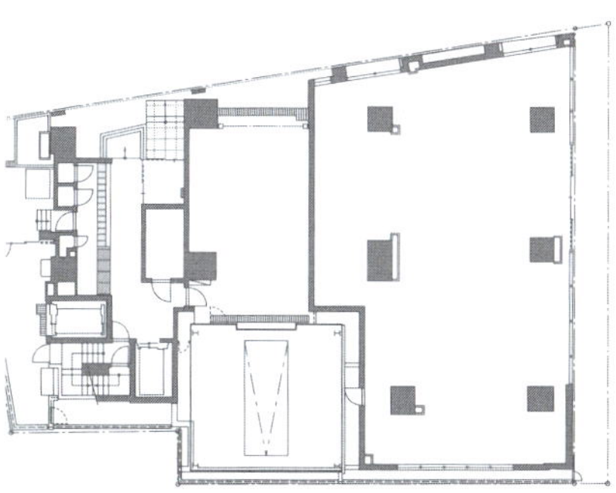

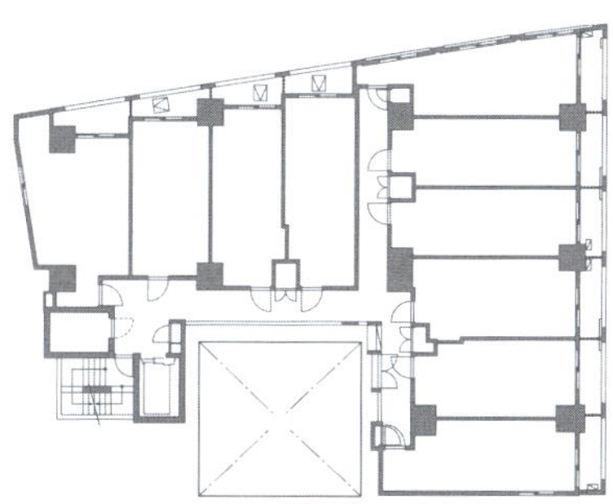

Street level views of the building highlighting the grid like façade and the brickwork.

Floor Plan.

JONAS

AMSTERDAM, THE NETHERLANDS

Jonas is a sustainable housing project in Amsterdam IJburg. The project, inspired by local themes of water, quayside, and shipbuilding, features a timber pavilion with steps, a green exterior, a rooftop beach, and a city beach. The building's irregular openings create a unique and innovative façade. The interior is warm, welcoming, and spectacular, expressing a modern, sustainable community life. The building's spatial concept is based on traditional wooden ship construction and features a canyon inspired communal route, linking public and collective programs. Jonas also features commercial facilities and a zero-energy system with solar panels to provide a future-proof energy supply. The project reduces material and focuses on reuse and on environmental impact. Jonas uses recycled materials and nature-inclusive landscaping to strengthen ecological values. Therefore Jonas has been awarded the BREEAM outstanding certificate. The building is designed to stimulate social interaction and connect collective spaces, thus creating a great living environment.

Building and reflection at Krijn Taconiskade.

ARCHITECT: ORANGE ARCHITECTS /// **COMPLETION:** 2022 /// **CLIENT:** AMVEST /// **INVESTOR:** PRIVATE /// **LANDSCAPE DESIGN:** FELIXX LANDSCAPE ARCHITECTS /// **INTERIOR DESIGN:** PUBBLIK&VOS /// **CONTEXT:** URBAN /// **GROSS FLOOR AREA:** 29,950 M^2 /// **NUMBER OF APARTMENTS:** 273 /// **MAIN SUSTAINABLE FEATURES:** CONCRETE STRUCTURE WITH RECYCLED AGGREGATES, COOLING WITH AQUATHERMAL ENERGY, BREEAM OUTSTANDING /// **PHOTOS:** SEBASTIAN VAN DAMME / WWW.SEBASTIANVANDAMME.NL (126, 128 B.R.) / STIJN POELSTRA / WWW.STIJNPOELSTRA.COM (127, 128 A., 129 A. L., 130-131) / EMILE HOENS / FLARE DEPARTMENT / WWW.FLAREDEPARTMENT.COM (129, B.)

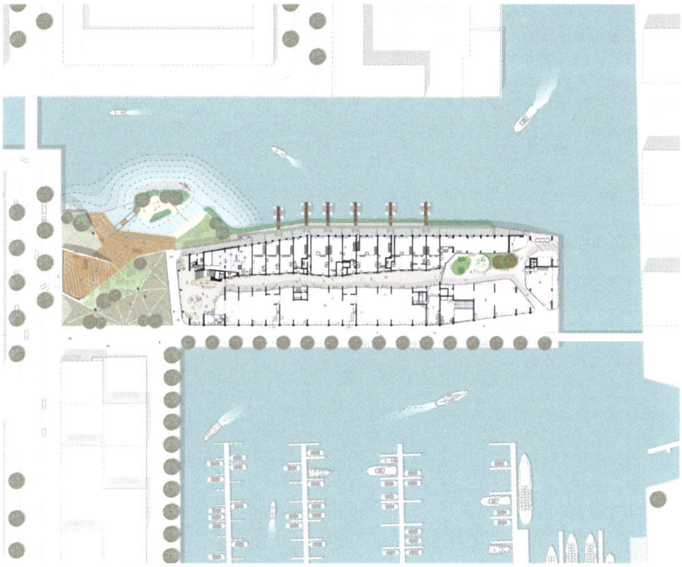

Site plan.

Front view of Jonas from the docks, as well as a view of a shared space inside the building.

A view of the cinema in the building as well as a view of the canyon-like passage inside.

Cross section.

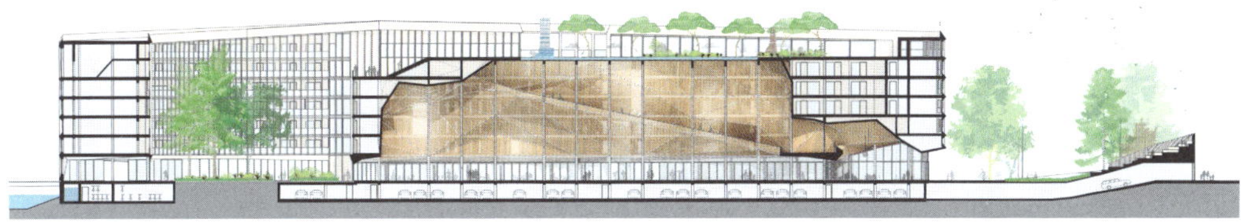

A view of the internal courtyard and the rooftop garden.

Longitudinal section.

FACTORY

AMSTERDAM, NETHERLANDS

The former Fluor office building in Haarlem was designed in 1972 with the idea that it could accommodate any future function. Initially it housed offices, but now the cubicles have been replaced by loft apartments. Of the 96 lofts, 30 are fully adaptable. Kitchen, bathroom, and bedroom can be placed anywhere. Elephant designed the layout together with the future residents, resulting in 30 unique loft apartments, perfectly tailored to the individual living preferences of the residents. By placing the new façade outside the existing structure, the façade columns are left exposed, preserving the traces of the original building. Large glass bays, stretching from floor to ceiling, bring ample daylight deep into the living space. On the ground floor, the windows fold inwards to provide shelter and create an outdoor space. From the first floor, the façade continues to fold outwards, adding extra space to the home. Symmetrical and asymmetrical bays add to the versatility of the layout, while the exterior plays with an interaction of transparency and reflection.

A view of the building from ground level.

ARCHITECTS: ELEPHANT /// **COMPLETION:** 2023 /// **CLIENT:** DURA VERMEER /// **INVESTOR:** PRIVATE /// **LANDSCAPE DESIGN:** ZUS /// **ORIGINAL BUILDING:** OFFICE BUILDING, 1972 /// **CONTEXT:** SUBURBAN /// **GROSS FLOOR AREA:** 18,000 M^2 /// **NUMBER OF APARTMENTS:** 96 /// **PHOTOS:** EMILE HOENS / FLARE DEPARTMENT / WWW.FLAREDEPARTMENT.COM

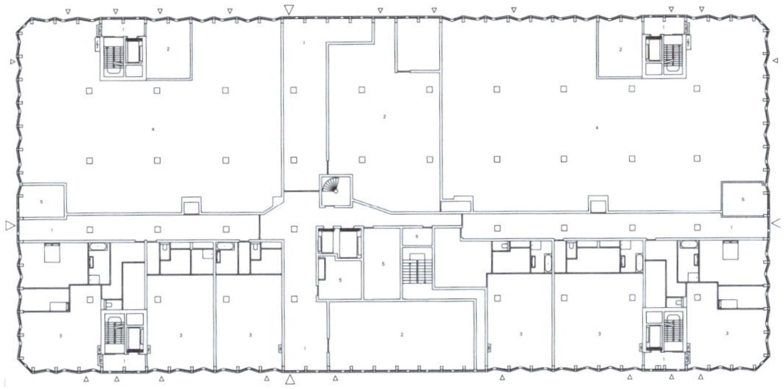

Ground floor plan.

Detailed close up of the unique glass façade of the building.

Various views providing examples of kitchens and living rooms in the individually designed apartments.

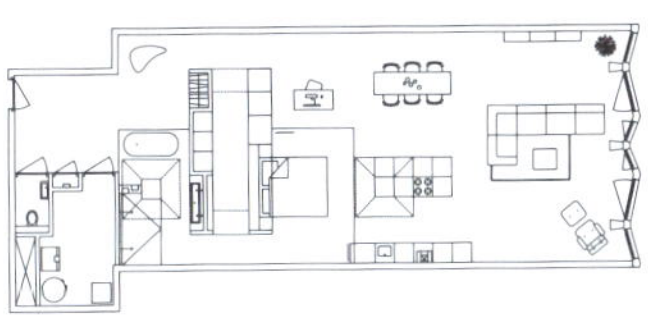

Different perspectives of the exterior of the building, highlighting the façade and the roof garden.

Typical room plan.

ORDYNKA

MOSCOW, RUSSIA

The project is located on the site of a famous chocolate factory in the heart of Moscow. It comprises seven high-end residential buildings connected by a spacious courtyard, which is unique for this very dense part of the city. The original façades of the factory have been renovated to preserve its historic appearance. In contrast, the façades facing the courtyard have a strikingly modern design. These elements give the building a distinctly modern character, with polylines, panoramic windows, and large modules of different materials.
The apartment building Ordynka also features a carefully restored 19th century merchant mansion. The original façade faces Malaya Ordynka Street and hides the inner courtyard formed by two linear volumes with façades connected by two raised bridges. The façades are clad in contrasting shades of light stone and dark metal. All apartment entrances face the focal point of the courtyard. The restored rear façade of the historic mansion creates a sense of place and community for residents.

Spacious courtyard connecting the buildings.

Views of Ordynkas unique façade from inside the courtyard.

Site Plan.

ARCHITECTS: AI STUDIO /// **COMPLETION:** 2021 /// **CLIENT:** INSIGMA /// **LANDSCAPE DESIGN:** ALPHABET CITY & NOVASCAPE /// **ORIGINAL BUILDING:** CHOCOLATE FACTORY, LATE 19TH CENTURY /// **CONTEXT:** CITY CENTER /// **GROSS FLOOR AREA:** 18,300 M² /// **NUMBER OF APARTMENTS:** 82 /// **PHOTOS:** DMITRIY YAKOVLEV (139, A.L.), DMITRIY CHEBANENKO/ WWW.CHEBANENKO.COM

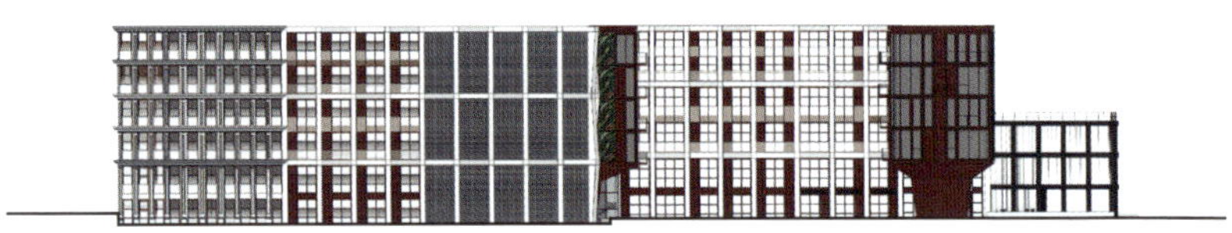

Close up view on the courtyard façade highlighting the contrast between the two styles.

Courtyard façade sketch.

The internal courtyard, the spectacular façade and the street view of the historic factory building.

A street-level view of the exterior of Brisbane 1

ARCHITECTS: BUREAU^PROBERTS /// **COMPLETION:** 2020 /// **CLIENT:** R&F PROPERTY AUSTRALIA /// **INVESTOR:** PRIVATE /// **LANDSCAPE DESIGN:** PLACE DESIGN GROUP /// **CONTEXT:** CITY CENTER /// **GROSS FLOOR AREA:** 95,300 M^2 /// **NUMBER OF APARTMENTS:** 608 /// **PHOTOS:** CHRISTOPHER FREDERIK JONES / WWW.CFJPHOTO.COM.AU

BRISBANE 1

BRISBANE, AUSTRALIA

Brisbane 1 is strongly influenced by its siting on a triangular block and its relationship to the nearby Brisbane River. The architecture is formed by the imagining of a single stone mass, eroded by water to create three distinctive towers. The ambition to resolve the building's complex shape is rewarded by long-range views which impart a shifting perspective and reveal, at distance, a crystalline form. The subtle manipulation of the floor plate heightens the intensity of the building's shape-shifting mass. Sun hoods add depth to the whole, bringing another layer of complexity and sense of movement to the façade. Verdant communal gardens provide outdoor gathering spaces and overhanging landscape offers passers-by a glimpse of this natural oasis, bringing a sense of theatre to the streetscape. At rooftop level, additional social spaces reveal new perspectives of the river city.

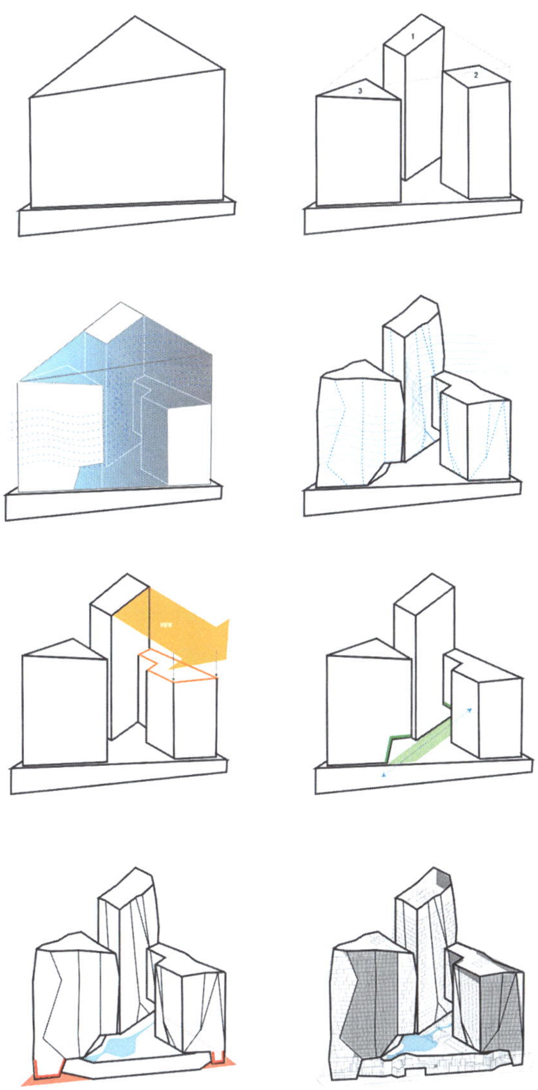

A view of Brisbane 1's sculpted form from the river.

Brisbane 1 Diagram

Brisbane 1 in the cityscape and its exterior and interior features and details.

Brisbane 1 as it rises above the river and a view of its pool area.

A view over the green roof of the building towards the north of Paris.

Longitudinal section.

ARCHITECTS: STUDIO RAVAZI + PARTNERS /// **COMPLETION:** 2023 /// **CLIENT:** EMERIGE, BROWNFIELDS, EXETER, FRANCE LOGIS /// **INVESTOR:** PRIVATE /// **LANDSCAPE DESIGN:** OLM /// **ENVIRONMENTAL DESIGN:** OASIIS /// **CONTEXT:** URBAN /// **GROSS FLOOR AREA:** 10,586 M² /// **NUMBER OF APARTMENTS:** 168 /// **MAIN SUSTAINABLE FEATURES:** GREEN ROOF, THERMAL INSULATION, NATURAL CROSS VENTILATION/// **PHOTOS:** SCHNEPP RENOU / WWW.SCHNEPP-RENOU.COM

93 PETIT

PARIS, FRANCE

93 Petit aims to be a building that addresses and attempts to solve problems currently faced by modern architecture in urban contexts. Large building sites and projects put pressure on the architect to find a sensible balance between land use and design. At present, urban design often involves the densification of the urban area, allowing a greater number of people to access public transport and city centers, as well as creating dense, lively and vibrant neighborhoods. Architecture office Ravazi + Partners aimed to design a building that would keep current ideas of urban design, while at the same time creating a clear morphological relationship with the adjacent urban fabric through massing, courtyards, choice of materials, fenestration, and size. The buildings light colored brick façade, green roof, and structured exterior design give it a modern look, that fits in seamlessly into the urban area of Paris while also adding sustainable features.

Overall view of main façade on rue Petit and the brickwork on rue Petit as well as the solid wood window frames.

The inner courtyard, a street level view, and the entrance area.

Close-up of the balconies and view of one of the balconies with built-in planting space.

Transverse cross section.

THE JAGGER

LOS ANGELES, CA, USA

On a rapidly changing thoroughfare in West Los Angeles, this mixed-use building carves a deep courtyard oasis into a bold and striking outline. Wrapped in a monolithic and relentless perforated window pattern, the building seeks to embrace the scale of higher density zoning along major boulevards. The building is topped with a sawtooth roof that creates an iconic silhouette. Two of the sloping roofs are made of wire to allow light into the stairwell below. The Jagger maintains a human-scale presence on the sidewalk with delicate green walls on the ground floor. An atrium leads to 74 apartments on six floors. Filled with a variety of plants it contributes to a pleasant and calming atmosphere. A variety of shared open spaces on multiple levels and an active and vibrant circulation system contribute to both communal and individual living.

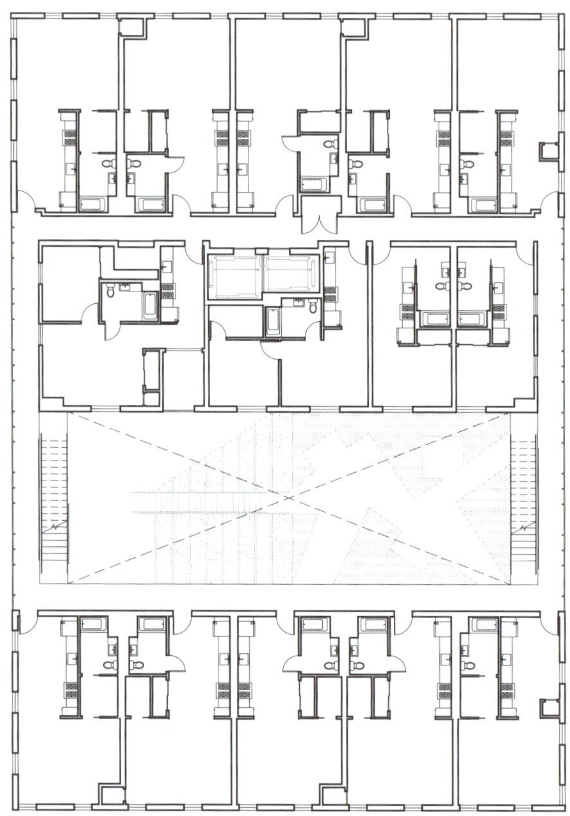

Side elevation highlighting window pattern and view looking into the entry courtyard from the street.

Ground floor plan.

ARCHITECTS: HOPSON RODSTROM DESIGN /// **COMPLETION:** 2023 /// **CLIENT:** CALE ENTERPRISES /// **INVESTOR:** PRIVATE /// **LANDSCAPE DESIGN:** TINA CHEE LANDSCAPE STUDIO /// **CONTEXT:** URBAN /// **GROSS FLOOR AREA:** 8,300 M² /// **NUMBER OF APARTMENTS:** 74 /// **PHOTOS:** PAUL VU / WWW.HEREANDNOWAGENCY.COM

Interior view of typical unit and view from street level showing surrounding context.

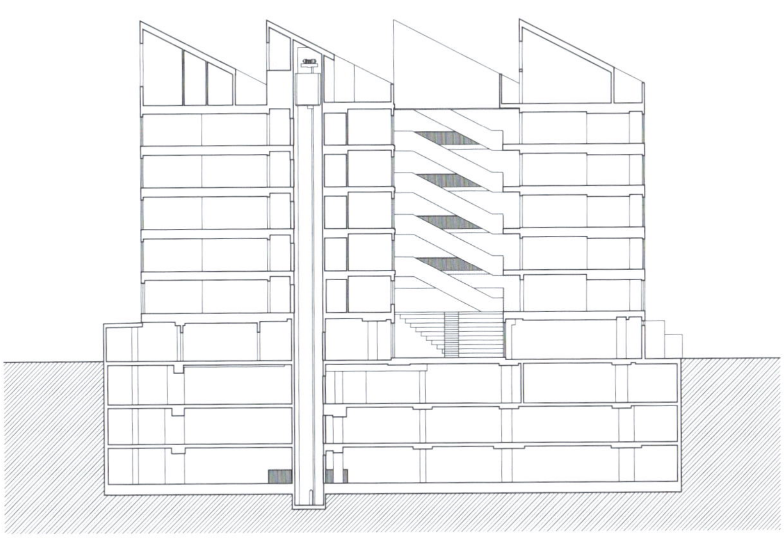

Exterior circulation close-up and courtyard stair view.

Cross section.

A view of the building curving around the center.

Cross section.

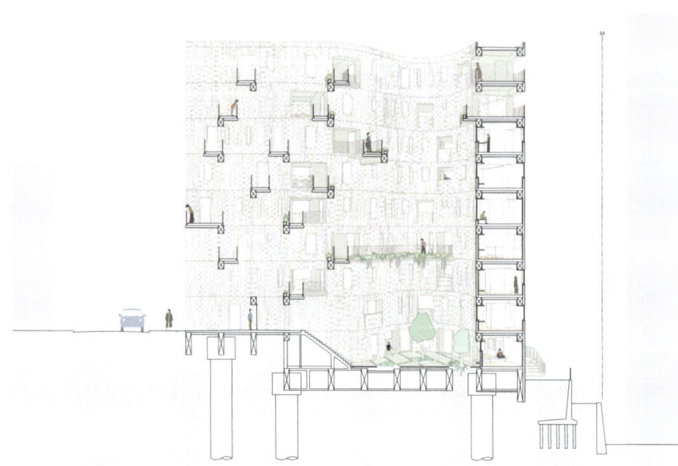

ARCHITECTS: HIROYUKI ITO ARCHITECTS /// **COMPLETION:** 2023 /// **CONTEXT:** CITY CENTER /// **GROSS FLOOR AREA:** 2,449 M² /// **NUMBER OF APARTMENTS:** 35 /// **PHOTOS:** MASAO NISHIKAWA / WWW.MASAONISHIKAWA.COM

TENJINCHO PLACE

TOKYO, JAPAN

The site located in Yushima, Tokyo is surrounded by high-rise buildings on three sides. The shape of the site and the light conditions were key factors in the design. By wrapping a curved volume around a central void, each unit receives light, wind, and views. To brighten up the dim center of the building and make it more inviting, three key strategies were introduced to create a brighter space. First, the few corridors that were built are converted into maisonette units. Second, openings to the courtyard were created as balconies to provide light and airflow. Thirdly, a texture was applied to the walls to emphasize the available light. By using unconventional materials, the aim was to create a linear yet irregular appearance. The courtyard formwork is made of Japanese cedar, cut into thin pieces with the bark on one side and attached to plywood, capturing the shapes and contours of the wood while playing with light and shadow. The courtyard is designed to become a public space in the future.

View from the street and within the courtyard, showcasing the connecting walkways and the wooden formwork.

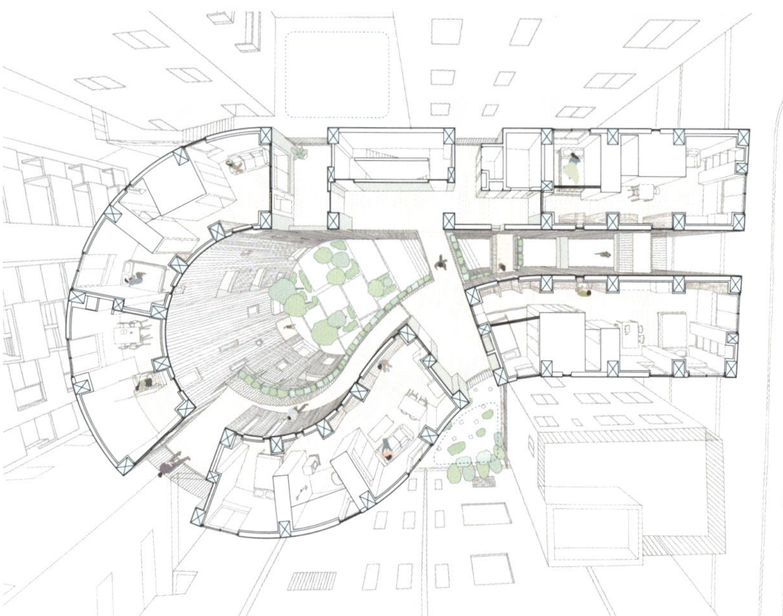

Views of within the building's apartments, highlighting the corridors following the building's curve.

Vertical perspective section.

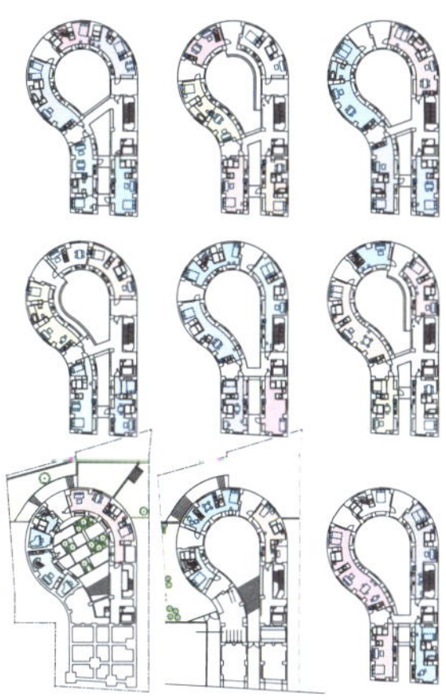

The interior courtyard walkways, balconies, and open spaces.

Floor plans.

STORIES

AMSTERDAM, THE NETHERLANDS

The Stories apartment building is constructed largely from negative carbon wood and uses open construction principles to ensure longevity. The 45-meter green tower addresses the challenge of communal, high density, sustainable and healthy urban living. The building combines 29 apartments and co-working units, a common area with gym, sauna, kitchen and outdoor urban farming, six commercial spaces in the base, and an underground parking garage. Characteristic of Stories' appearance is its deep white steel façade, which contains extensive outdoor spaces in the form of private balconies and 57 double height vegetation units, which also act as architectural shading elements with a cooling effect on the building. The larger of these are niches for various types of trees, shrubs, grasses and associated fauna, creating vertical greenery. They act as privacy screens between apartments and provide the infrastructure for a healthy microclimate, enriching the local urban biodiversity and giving seasonal expression to the design.

White steel façade with private balconies and double height vegetation units.

A front and side view of the building's unique façade as seen from the canal.

A great variety of floor plans and housing typologies to suit diverse needs.

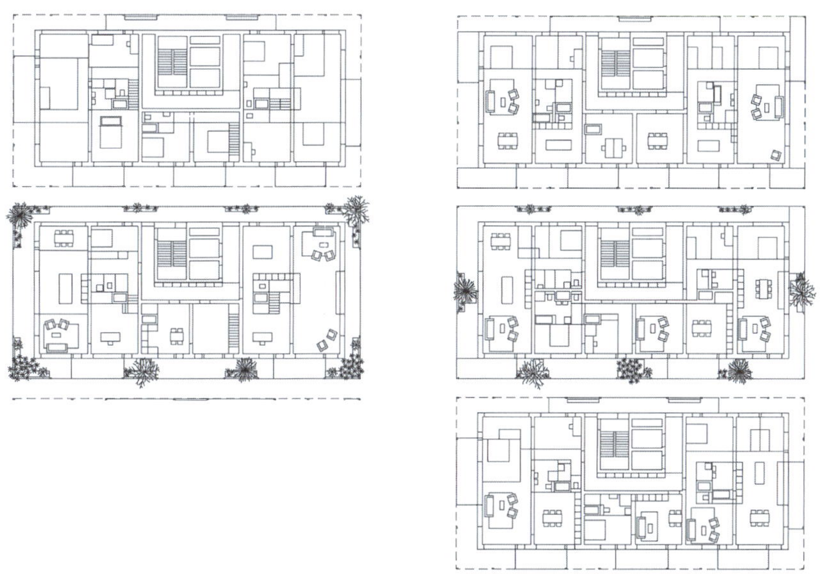

ARCHITECTS: OLAF GIPSER ARCHITECTS /// **COMPLETION:** 2021 /// **CLIENT:** SELF-BUILT COOPERATIVE BSH20A HEUTINK GROEP /// **CONTRACTOR:** HEUTINK GROEP /// **LANDSCAPE DESIGN:** SMARTLAND LANDSCAPE ARCHITECTS /// **MAIN STRUCTURAL ENGINEERING:** ALFERINK VAN SCHIEVEEN /// **CONTEXT:** URBAN /// **GROSS FLOOR AREA:** 5,500 M² /// **NUMBER OF APARTMENTS:** 29 /// **MAIN SUSTAINABLE FEATURES:** RENEWABLE RESOURCES, LOW CO2 FOOTPRINT, POST-ANTHROPOCENTRIC ARCHITECTURAL CONCEPT WITH DEEP FAÇADE, OPEN BUILDING PRINCIPLES, SOCIAL SUSTAINABILITY /// **PHOTOS:** LUUK KRAMER/WWW.LUUKKRAMER.NL (161 A.L.), MAX HART NIBBRIG/ WWW.MWAHARTNIBBRIG.COM

An inviting outdoor space enriched by double height greenery. Custom designed apartment interior featuring an exposed cross-laminated timber structure in the ceiling.

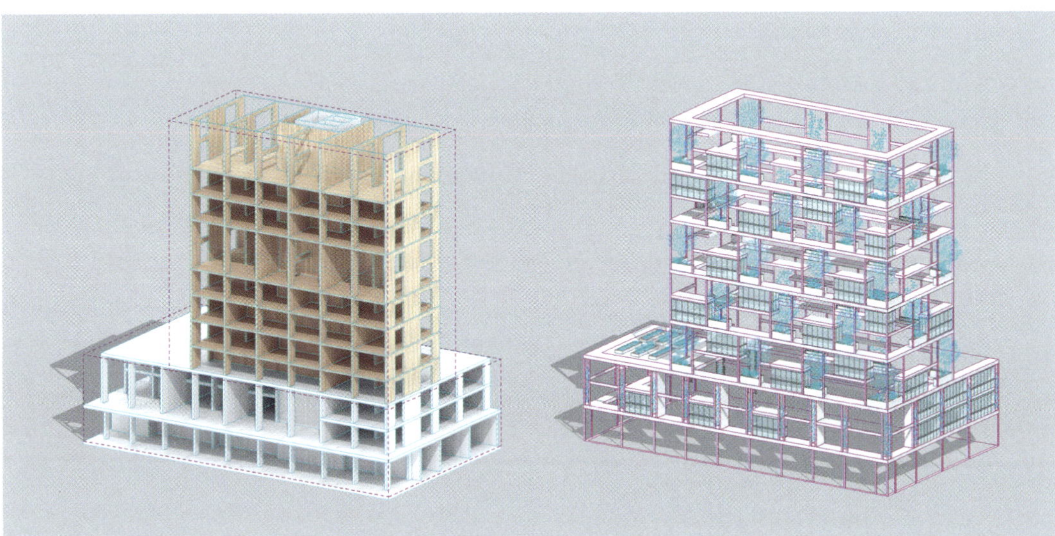

The double height niches with vertical greenery create a great harmony of architecture and nature.

Diagrams of the hybrid wooden structure and an enveloping steel framework with extensive outdoor spaces.

URBAN MONOLITH

BÈGLES, FRANCE

This architectural project, located on a complex plot in the Terres Neuves quarter of Bègles, stands out for its seamless integration into a dense and diverse urban environment. The monolithic white concrete building adapts to its context by creating unique façades for each orientation, maximizing natural light and enhancing the residents' quality of life.

The apartments are designed with bright living areas and quiet sleeping spaces, complimented by generously sized, well-oriented metal balconies. The slight shift in the position of the balconies gives a more textured feel to the façade of the monolith. Attention to durable materials and outdoor spaces reflects a strong environmental approach, while naturally lit, internal circulation areas encourage social interaction. The project plays a key role in transitioning between new developments and existing structures, ensuring a smooth, functional urban connection.

Building profile with attic in stained concrete and metal balconies.

Inflection of the north façade of the white cast-in-place concrete monolith.

ARCHITECTS: AUP ARCHITECTES /// **COMPLETION:** 2020 /// **CLIENT:** ATARAXIA FOR LA FONCIÈRE LOGEMENT /// **INVESTOR:** PUBLIC /// **CONTEXT:** URBAN /// **GROSS FLOOR AREA:** 2,536 M² /// **NUMBER OF APARTMENTS:** 31 /// **PHOTOS:** ANTOINE MERCUSOT /WWW.ANTOINEMERCUSOT.COM

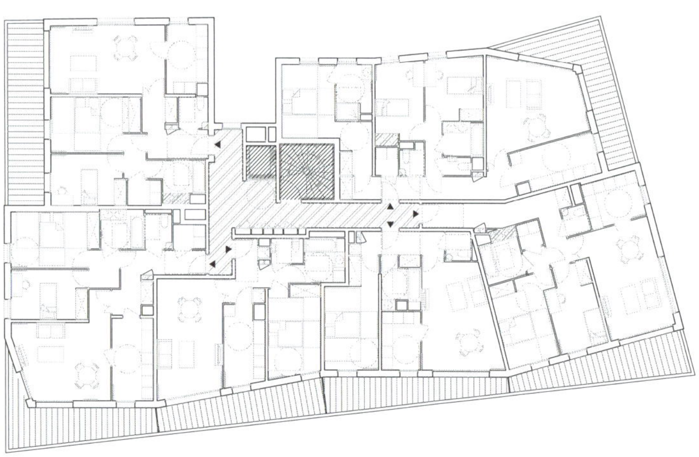

View of the northern entrance, where the volumes are offset to create a slender urban composition and a closer view of the building's contemporary layout: columns for the base, balconies for the upper floors and attics for the crown.

Ground floor plan.

Cross section.

Partial view of the south façade, punctuated by long balconies and attics in stained concrete.

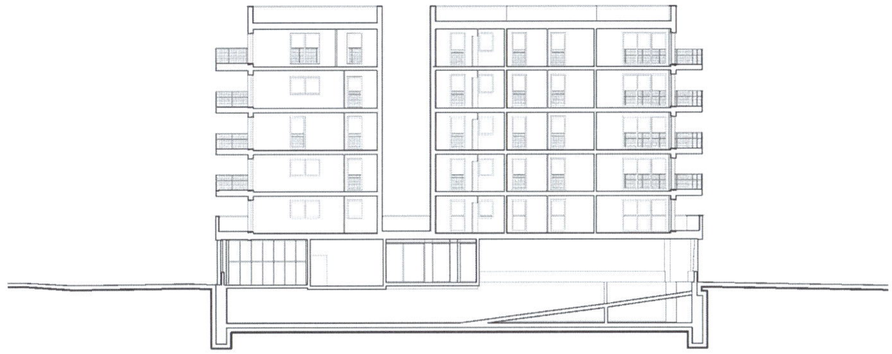

NORFOLK

BURLEIGH HEADS, AUSTRALIA

Drawing inspiration from local Norfolk Pines, this multi-residential project by Koichi Takada Architects is a sculptural and sympathetic addition to Burleigh Heads in Queensland, Australia. The use of the pine cone as inspiration was guided by the fact that in nature, pine cones protect their seeds from all kinds of adversity. Like the outer layers of the pine cone, the building provides its residents with a shelter and a retreat from the subtropical sun. Norfolk benefits from a northerly aspect on the 1,012 square-meter beachfront site at the southern end of Burleigh Heads Beach. The ten-story structure provides 15 apartments, including two dual level penthouses with private rooftop pools and impressive ground floor wellbeing amenities that allow residents to enjoy the prime location year-round, including a gym, outdoor pool, and sauna. Overlapping tapered balcony slabs and sliding slatted screening are iconic components of this responsive building that has implemented passive design principles, resulting in a building that breathes.

The building's appearance is informed by the beachside surroundings. Curved horizontal battens form a central spine of this sculptural building and accentuate the organic expression of the design.

ARCHITECTS: KOICHI TAKADA ARCHITECTS /// **COMPLETION:** 2021 /// **CLIENT:** FORME /// **INVESTOR:** PRIVATE /// **INTERIOR DESIGN:** MIM DESIGN /// **CONTEXT:** COASTAL SUBURBAN /// **GROSS FLOOR AREA:** 8,510 M² /// **NUMBER OF APARTMENTS:** 15 /// **MAIN SUSTAINABLE FEATURES:** PASSIVE DESIGN PRINCIPLES, SITE RESPONSIVE SOLUTIONS /// **PHOTOS:** SCOTT BURROWS / WWW.SCOTTBURROWSPHOTOGRAPHER.COM (166, 167, 168B., 169B.) / TOM FERGUSON / WWW.TOMFERGUSON.COM.AU

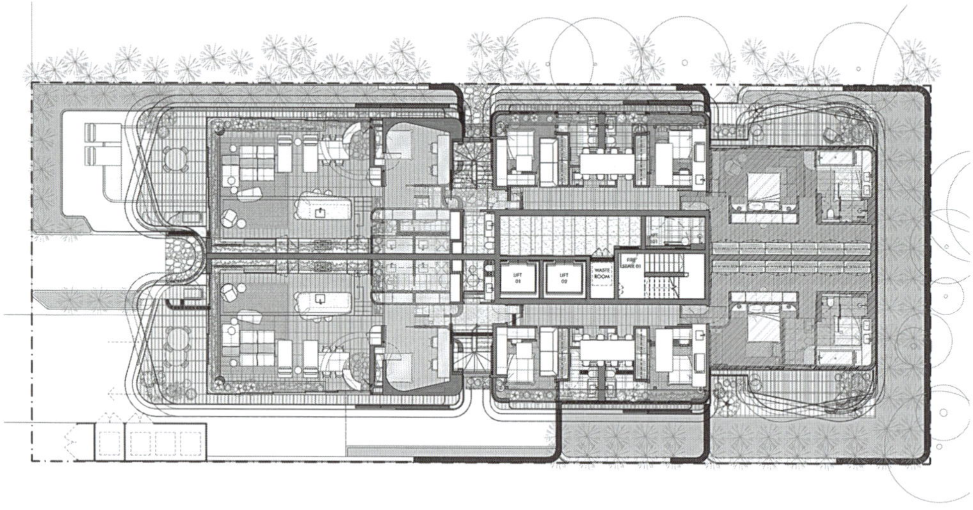

Typical floor plan.

The interior layout directs lines of sight towards the view. The apartments can be completely opened to allow fresh air to enter. Norfolk is located directly at Burleigh Heads Beach with views towards the scenically pleasing Gold Coast.

Inside a typical apartment and the flowing forms of the exterior. The façade, inspired in its appearance by the pine cone, interacts organically with the beach environment.

BOGEN

HUNINGUE, FRANCE

The 16-story Bogen Tower stands on the Rhine just like a lighthouse facing the river. With its architectural bias made of vertical strata, and its balconies whose railings are made of glass, the high-standing building attracts the eye with its sculptural style. Due to the building's lines and heights, the inhabited rooms benefit from views over the Rhine, the town, and the territory. Appearing like a signal tower from the Place Abbatucci in Huningue, the Three Countries Bridge naturally leads to it. The side façade accompanies pedestrians from the city center to the banks, while its largely glazed ground floor accommodates various shops that complete the shopping area and contribute to the creation of many places of living and sharing. Les Jetées district has been designed to allow the residents to enjoy the qualities of this exceptional site. All the accommodations benefit from a private outdoor space, largely vegetated and designed as a loggia to preserve the privacy of the residents. Thus, the offset of the tower blades offers an unobstructed view to every dwelling.

View of the building from the river.

Ground level shot of the building, showcasing its spectacular vertical strata design.

ARCHITECTS: PIETRIARCHITECTES /// **COMPLETION:** 2023 /// **CLIENT:** CONSTRUCTA PROMOTION /// **INVESTOR:** PRIVATE /// **LANDSCAPE DESIGN:** LAND'ACT /// **CONTEXT:** URBAN /// **GROSS FLOOR AREA:** 5,319 M^2 /// **NUMBER OF APARTMENTS:** 71 /// **PHOTOS:** LUC BOEGLY

View from within an apartment as well as a close up of the building's façade and balconies.

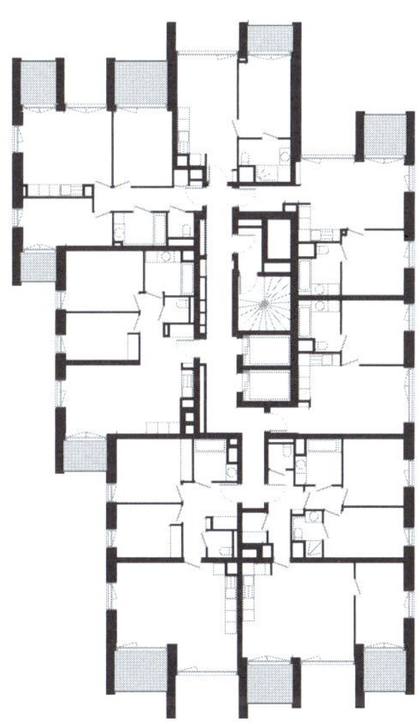

Views of the building's white metal façade from different perspectives.

Typical floor plan.

A street-level view of LIVE EUR that contextualizes its volume within its surroundings.

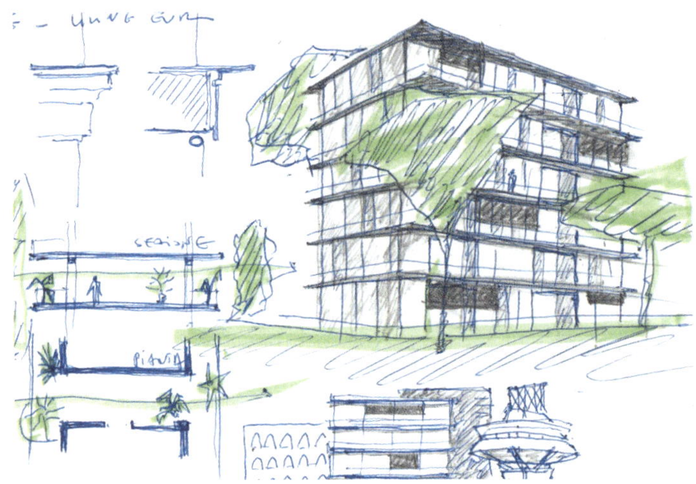

Design sketch of LIVE EUR.

ARCHITECTS: IT'S /// **COMPLETION:** 2023 /// **CLIENT:** BMV IMMOBILIARE S.R.L. /// **CONTEXT:** URBAN /// **GROSS FLOOR AREA:** 1,600 M^2 /// **NUMBER OF APARTMENTS:** 12 /// **MAIN SUSTAINABLE FEATURES:** HEAT PUMPS, SOLAR PANELS, DOMOTICS, AUTOMATIC INTERNAL TEMPERATURE REGULATION/// **PHOTOS:** FRANCESCO MATTUZZI (178 B.), FRANCESCA IOVENE / WWW.FRANCESCAIOVENE.COM

LIVE EUR

ROME, ITALY

LIVE EUR is a new residential building in Rome whose design was inspired by the main principles of the Active House: environmental sustainability, energy efficiency, and comfort. The building consists of seven floors above ground, 12 apartments, and a penthouse. The architects' choice of high quality materials and construction techniques results in comfortable and highly energy efficient interiors. The building is powered by heat pumps and solar panels, which are renewable energy sources that make the building environmentally neutral.
All apartments have openings on two opposite sides and all internal layouts include a living area facing both sides. The orientation of the building maximizes the amount of natural light and ventilation. Realized in reinforced concrete, with a system of prefabricated and disposable formwork, the building presents a simple and clean volume as a homage to the best examples of italian architecture of the sixties and seventies.

A front view of the building hightlighting its three part structure and a close-up of the balconies, showing the building's openness to natural light.

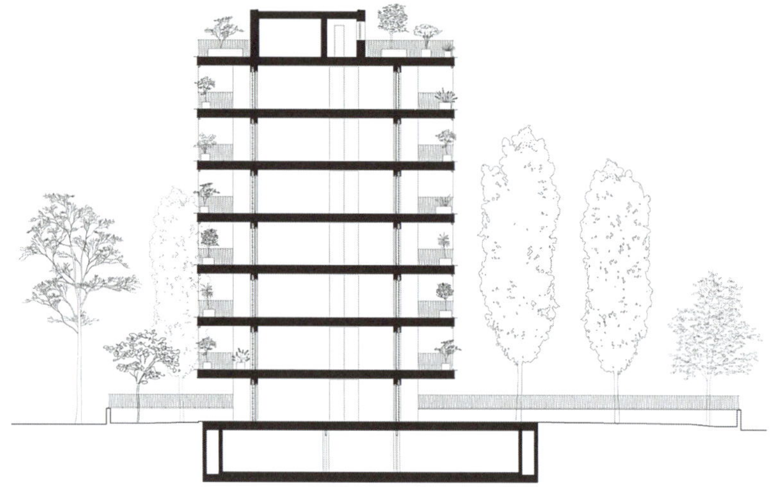

Views from the inside of one of the apartments and the entrance area.

Cross section.

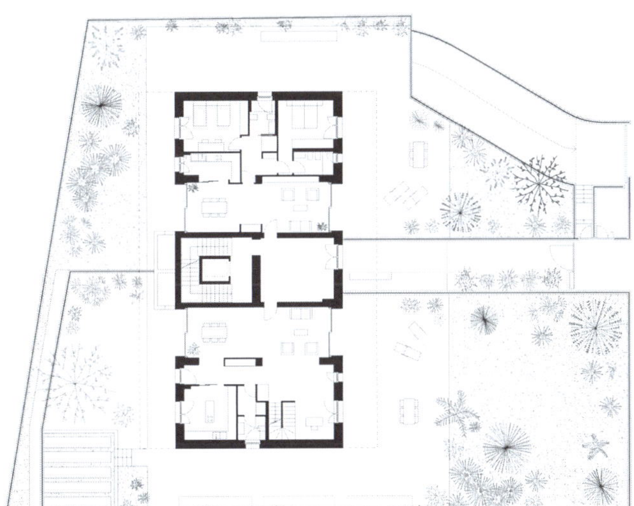

The travestine interior of the building and a view of LIVE EUR's balconies and façade.

Ground floor plan.

LE STELLA

MONACO, MONACO

The uniqueness of the project lies in its abstract form. Because of this abstraction and simplicity, the building is not perceived as a static mass, but appears to be in motion. The curves and bends of the façade offer passers-by ever-changing perspectives of the project. The spectacular façade plays with light, emphasizing the aspect of the building as an urban sculpture. This is supported by the large white façade that surrounds and protects the building. Inspired by a sail and covered in white mosaic, Le Stella creates its own landscape and atmosphere. The building houses imposing loggias that block the view of the neighboring buildings while allowing a wide view of the sea. The project has 16 floors and is divided horizontally into two volumes. The building not only contains several residential units, but also a glass base that houses official spaces such as shops, offices, and a school. A key element of the building are the glass balustraded passageways that connect the two volumes. These suspended walkways allow and encourage more communication among residents.

Curves and bends make the façade appear to be in motion.

Close up of the spectacular façade and a ground shot of the building.

ARCHITECTS: JEAN-PIERRE LOTT – ARCHITECTE /// **COMPLETION:** 2018 /// **CLIENT:** JB PASTOR & FILS /// **INVESTOR:** PRIVATE /// **CONTEXT:** CITY CENTER /// **GROSS FLOOR AREA:** 15,000 M^2 /// **NUMBER OF APARTMENTS:** 99 /// **MAIN SUSTAINABLE FEATURES:** BÂTIMENT DURABLE MÉDITERRANÉEN – NIVEAU OR /// **PHOTOS:** ALDO AMORETTI / WWW.ALDOAMORETTI.COM

Various detailed views of the suspended walkways that connect the buildings, and the view of the harbor from the terrace.

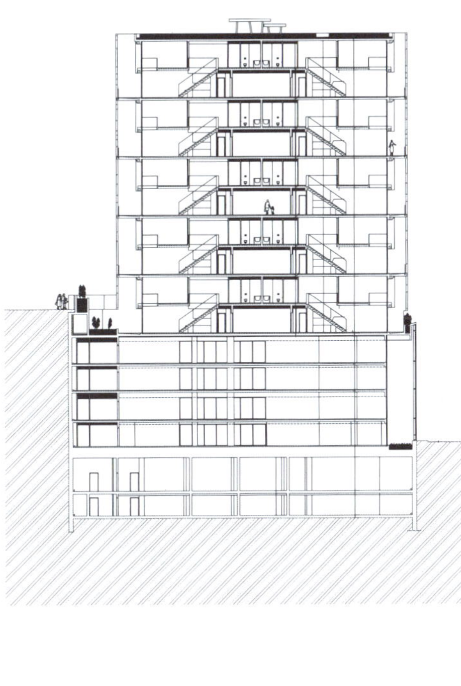

Inside a typical apartment.

Section of one of the two volumes.

KAKTUS TOWERS

COPENHAGEN, DENMARK

The Kaktus Towers, designed by BIG – Bjarke Ingels Group, are two iconic and unique apartment buildings in the Copenhagen cityscape. The towers feature a repetitive structure that is rotated slightly differently on each level, creating a spectacular look that, as the name suggests, resembles a cactus plant with its spiky exterior. The twisting terraces also provide generous balconies for the apartments. In addition, the two high-rise buildings are connected by a sloping public park that extends from the roof of a nearby commercial building towards the towers. This creates a pleasant view and atmosphere while enhancing the quality of life in the neighborhood. The buildings also provide access to an outdoor kitchen, barbecue area, café, various fitness rooms and a party room. Kaktus Towers is also located close to the city center, providing residents with many nearby options. The interior of the buildings is simple yet aesthetic, providing a calm counterbalance to the striking exterior of the towers.

Frontal view of the Kaktus Towers from the street.

Close up view of the balconies and twisted terraces of one of the buildings.

ARCHITECTS: BIG – BJARKE INGELS GROUP /// **COMPLETION:** 2024 /// **CLIENT:** KAKTUS 1 PROPCO APS V/CATELLA /// **INVESTOR:** PRIVATE /// **CONTEXT:** CITY CENTER /// **GROSS FLOOR AREA:** 26,100 M² /// **NUMBER OF APARTMENTS:** 495 /// **PHOTOS:** RASMUS HJORTSHØJ / WWW.RASMUSHJORTSHOJ.COM

Views of a typical apartment#s interior and a corridor in the tower.

View of the differently twisted and angled terraces that create an iconic spiky exterior and view from a balcony.

Cross Section.

12KR

INDEX
ARCHITECTS

INDEX
PHOTOGRAPHERS

IMPRINT

The Deutsche Nationalbibliothek lists this publication in the Deutsche Nationalbibliografie; detailed bibliographic data are available on the Internet at http://dnb.dnb.de.

ISBN 978-3-03768-302-6

www.braun–publishing.ch

1st edition 2025

All of the information in this volume has been compiled to the best of the editor's knowledge. It is based on the information provided to the publisher by the architects' and designers' offices and excludes any liability. The publisher assumes no responsibility for its accuracy or completeness as well as copyright discrepancies and refers to the specified sources (architects' and designers' offices). All rights to the photographs are property of the photographer (please refer to the picture credits).

Editor
Editorial Office van Uffelen

Editorial staff and layout
Alexander Peschka, Chris van Uffelen

Graphic concept
Michaela Prinz, Berlin

Reproduction
Bild1Druck GmbH, Berlin

All of the information in this volume has been compiled to the best of the editor's knowledge. It is based on the information provided to the publisher by the architects' and designers' offices and excludes any liability. The publisher assumes no responsibility for its accuracy or completeness as well as copyright discrepancies and refers to the specified sources (architects' and designers' offices). All rights to the photographs are property of the photographer (please refer to the pages).

Product Safety
Publisher: Braun Publishing AG, Arenenbergstr. 2, 8268 Salenstein, Switzerland, publisher@braun-publishing.ch
EU-Representative: Bookwise GmbH, Zeppelinstr. 67, 81669 Munich, Germany, info@bookwise.de